CW01430177

Blessings!
Kim Carlsberg

The Art Of Close Encounters

PRAISES FOR:
THE ART OF CLOSE ENCOUNTERS...

Visually stunning, intellectually compelling
and spiritually illuminating, **The Art Of Close
Encounters** provides the reader with a treasury
of the variety of contact between humans,
extraterrestrial and other non-human entities.
As editor and coordinator of the listed productions,
and a researcher who seeks to confirm the reports
behind the images, Kim is to be commended.
The result of her work expands our awareness
of our cosmic cousins. May we experience a shift
in consciousness from Planetary Persons, to
Cosmic Citizens.
–R. Leo Sprinkle, PhD
Professor Emeritus, Counseling Services,
University of Wyoming

It pleases me to contribute to such a vital anthol-
ogy of UFO related illustrations. Human interaction
with extraterrestrial / extradimensional entities
is such a profoundly visceral experience. Dry,
written, documentary reports simply cannot
convey the awesome impact of these astounding
encounters. Kim's magnificent book magically
brings this enigma to life, and is an integral tool
and asset for serious UFO research.
–Jim Nichols
UFO Artist

Over the years I have had the honor to
provide a forum for countless contactees to
voice their incredible close encounter stories.
Now, author and abductee, Kim Carlsberg has
put a face to these life altering experiences in
a long overdue, powerful and artistic presentation, **The Art Of Close Encounters**. The
sheer volume of stories, the authenticity and
range of the encounters, and the beauty and
diversity of the art makes this new book the
closest we will have to a reference guide on
contact for quite some time.
Remarkable achievement!
–George Noory
Host of "Coast To Coast" am radio.

The number of contactees unable to talk
about what they have experienced is not
known. Who will speak for those who cannot
yet speak for themselves? In this one of a
kind book, Kim Carlsberg has reached out
to bridge the silence with art, letting the
universal language bring together those
caught between two worlds. Every contactee
should have this book in solidarity with all
others. Let the dialogue begin.
–Stephen Bassett
Paradigm Research Group

The Art Of Close Encounters
by
Kim Carlsberg

ᴄᴇᴘ

Close Encounters Publishing

Library of Congress Cataloging-in-Publication Data
Carlsberg, Kim 1955-
The Art Of Close Encounters: by the author of
Beyond My Wildest Dreams / Kim Carlsberg, Non-Fiction

Library of Congress Control Number: 2010912691

ISBN 978-1-4507-3268-0
1. ET Art   2. Aliens   3. Contact   4. Paranormal   5. UFOs
6. Extraterrestrials   7. ETs   8. Sightings   9. Unidentified Flying
Objects   10. Close Encounters   11. Alien Art   12. UFO Art
13. ET Art

I. Title

Copyright © 2010 Kim Carlsberg. All rights reserved. No part
of this book may be reproduced by any means or in any form
whatsoever without written permission from the publisher,
except for brief quotations embodied in literary articles or
reviews. Each story and art piece is copyright of individual
contributor.

First Edition   Printed in China   October 2010

    1   2   3   4   5   6   7   8   9   10

CEP
CloseEncountersPublishing.com
Compilation, Editing, Art Direction,
Graphic Design, & Book Jacket, by Kim Carlsberg
Front Cover Image: *Eliad* by Kesara
Back Cover Image: *Adam* by Dale Ziemianski
Author Photo by Elliott Easton

This is my personal invitation to you to participate in the next volume of The Art Of Close Encounters. The response to the first volume was so overwhelming, I started the second volume before the first one was completed. If you have an encounter story you would like to share, please visit www.closeencounterspublishing.com. Your contribution will have a rippling effect on raising the consciousness of our world.

Kim Carlsberg... The Art Of Close Encounters.

Acknowledgements - May unfathomable good fortune follow and sustain Steven Bassett in his relentless quest to facilitate disclosure in our lifetimes. Thank you Steven for your vision, generosity and tenacity, certainly in your vigilant quest for all humankind, but also in the invaluable contributions you have made, and continue to make in bringing The Art Of Close Encouters to the world consciousness.

... deepest regards to Duncan Roads of NEXUS Magazine for his special commitment to seeing that the news of the book is being spread throughout the globe.

... special blessings to Dennis Briefer and Sharon McCormick for their faith and assistance in seeing the book through to the end.

... my sincerest gratitude and love to Phil Christy. Your unconditional support, positive attitude and rich sense of humor were, and are, life enhancing.

... huge heart felt hugs to my girls; Elle Keith, Joliebeth Cope and Melissa Kriger, for your selfless generosity and love.

... enormous thanks to Larry Lowe for caring for my animals, for always having a level head, an engaging intellect, impeccable integrity and a warm and giving soul.

... cheers to my fantastic Facebook family who consistently make life interesting and fun. I greatly appreciate all of you for supporting who I am and what I believe to be important the way you do.

Dedications - This book is dedicated first and foremost to the contributors; for their stories, art, spirit, bravery, support and trust... after thousands of emails, I have a whole new world of absolutely amazing friends.

... to all abductees, contactees, experiencers, authors, researchers, members of the media, and everyone who has had the courage to believe and put forth the truth in the face of continued ignorance and ridicule... you are my heroes.

... to my mother Freida Riggin, for showing me what it is to be a strong woman in the world, to go where others fear to go, and to have faith in something.

... to my brother Chet Riggin.

... to my cat Bounce who kept me company through months upon months of all-nighters.

... to my precious dog Baby who rarely saw the beach the entire year.

... and to whom life has proven to be my greatest supporter and dearest friend, Darryl Anka.

Introduction - "The Art Of Close Encounters" came into being as a result of two subjects that are deeply personal to me... UFOs and art. I am an author on the subject of UFOs and ETs and I am a professional photographic artist.

I was born a rebellious child. When left alone I would paint murals on my bedroom walls. My mother would arrive home, throw a fit, have the room repainted and I would paint another mural.

Little did I know she was secretly photographing the murals each time. She obviously knew the best way to get me to do anything was to tell me I couldn't... so I continued to draw and paint.

I was offered a full scholarship to the Kansas City Art Institute, where I lived at the time, but I had already had my fill of Midwest winters, so I made my way to California.

I studied art and photography at Santa Monica College and eventually graduated from the Pasadena "Art Center College Of Design", in commercial photography.

I was accepted into Art Center by submission of both my illustration and photography portfolios. At that time there was no such thing as "a split major": I was forced to choose between illustration and photography, a pivotal moment in my life. Logic won out. I believed if times got bad, I could always make a living as a portrait photographer in any "Small Town" USA. In my mind, illustration jobs would certainly be harder to come by.

I always wondered if I made the right decision. During the first trimester of my photography course, one of the prerequisites for my major was a drawing class for photographers. It was a class I will never forget.

A few weeks into the trimester, the instructor whispered to me during session, "What on Earth are you doing in the photography department... you're an illustrator! If you were in this department, you would be one of the finest illustrators to ever graduate from this institution!"

I went home and thought it over again, and still, logic (fear most likely) prevailed. I stayed in the photography department and never picked up a pencil, pastel, or paint brush again... but I have never lost my passion for art.

I did my best with the choice I had made and became the first woman accepted into IATSE, Local 659, the Hollywood "camera-persons" union (then known as the "camera-mans" union). Life was grand. I was living on the beach in Malibu and working on the set of BayWatch when... "one ordinary night, in the middle of an ordinary life, I had an extraordinary dream", which changed my life forever.

Although, up until that time, I had never heard of "alien abduction", I suddenly found myself at the heart of the world of UFOs and extraterrestrial encounters. The contacts continued for years and I faithfully recorded them in my diary which eventually led to my first book "Beyond My Wildest Dreams."

Grasping at comprehending the phenomenon of alien abduction, ETs and UFOs requires radically expanding one's concepts of the nature of reality, and a lot of research.

I found while studying the field of UFOs and close encounters, other esoteric subjects intertwined; telepathy, hands-on healing, etc., and contact with energies and entities other than "gray aliens" were beginning to look commonplace.

I immersed myself into this new realm of knowledge and as a result discovered my own latent abilities. I skeptically enrolled in a channeling class, which turned out to be one of the best decisions I ever made.

I learned valuable skills and incorporated practices into my life that have been wonderfully enriching. I met many fascinating "enlightened" people.

One in particular, Darryl Anka, would eventually become my mentor, best friend, and healer of the wounds of abduction.

World renowned channel for the compassionate, wise (and humorous) human/gray hybrid, extraterrestrial entity known as "Bashar", Darryl's professional and creative talents are broad sweeping as well. He is a brilliant writer, director, producer, artist, and art-director, to mention a few.

Darryl and I spent a lot of time together while I was writing "Beyond My Wildest Dreams - Diary of a UFO Abductee." We are both extremely visual which largely influenced its presentation. I was honored when Darryl offered to graphically illustrate my diary.

Since the publication of that book, we have received countless compliments on the power the illustrations have had in conveying the overall story, rather than if the book had consisted of text alone.

Thus... the evolution of "The Art Of Close Encounters." My story has already been told, but scores of people have their own tales to tell. For most, they are not enough to fill a book, but combined with others' accounts... I believe volumes are yet to be filled.

About The Book - When compiling a project like this, one becomes extremely intimate with the contributors and the material. Especially when one has a skeleton staff... me.

The process starts with communicating with the writer as to the appropriateness of their story for the book.

Many people are initially shy about the value of their submission... "I had an encounter, but it was nothing really." Then, with a little prodding, electrifying and (usually long) accounts fall from their lips.

Thus creating the challenge of editing the material to the "one page" of text space reserved to support each piece of artwork.

The stories are edited, formatted and proof-read... re-edited, re-formatted and re-proofed, etc.

The contributors are rarely authors. Usually dozens of emails are exchanged, "is this what you are trying to say?"... "not exactly... it is more like this." Once the stories are relatively concise, they are categorized.

Many times accounts arrived unaccompanied by art. An artist must be solicited to interpret the story and render it accurately. Most times, that again, turns out to be me. Illustrating someone else's story requires sincere familiarity to insure an honest depiction of the event.

Finally, with so many great stories and art pieces submitted comes the agonizing process of which ones to exclude due to space limitation.

Next... the art itself. A lot of imagery today is produced on the computer and for the web, which creates a huge problem. Screen resolution for the web is extremely low, 72 DPI (dots per inch). Printing requires 300 DPI. Therefore, as a professional photographer, part of my arsenal is powerful enlarging software which is really getting a workout with the creation of this book.

Each piece of art must be "optimized"; sized, cropped, levels and/or contrast ratios adjusted - white points - black points, colors balanced, saturations modified, pixels sharpened, etc.

Lastly... the actual lay-out of the book. Formatting text is a painstaking and tedious task. Not my expertise by any means. Columns, paragraphs, hyphenations, dangling words, etc., are crazy making.

With this first edition of "The Art Of Close Encounters" I am working with unfamiliar, unsophisticated and extremely slow software.

Every command requires tedious amounts of time to process and system crashes are a regular occurrence. Words, sentences, paragraphs and sometimes entire pages simply vanish to be reconstructed.

I have learned a lot, and in the process, become so familiar with the stories, the art, and the contributors, they are all now - as old friends.

With the intimacy of this plethora of new information, I have been asked if I have come to any new conclusions about contact, and my answer is yes.

Oddly enough, they are not regarding the beings encountered. They are regarding the surprising magnanimity, bravery, openness and warmth of the human beings that have made this book possible, and the entirety of the "Encounter" communities in which I found them.

I surmise close encounters have probably remained quit similar throughout time. The change that is taking place now is occurring within us.

The baseline has been raised. The outdated question "Are we alone?" has given way to "Who are they?", "What are we to them?", and "When will we, as a global society, recognize and openly surrender to the undeniable fact that we are playing in an infinite and abundantly occupied field of existence.

I truly believed completing this book was a pipe dream. What has transpired has humbled me.

Instead of engaging in a process of "teeth pulling" which I fully anticipated, what I found was a brotherhood of regular folks, with honesty and generosity of spirit words cannot describe. Instead of meeting with fear and rejection, I was received with love and camaraderie.

Yes, much has changed since the early nineties when I wrote "Beyond My Wildest Dreams, Diary of a UFO Abductee." Thanks to the birth of the internet and the continuous work of researchers and courageous souls over the decades, the subjects of UFOs and encounters are no longer entirely taboo, but ones that now, more often than not, foster excitement, intrigue and the desire to share experiences and information.

I believe the internet is hugely responsible. The web not only offers opportunity for like-minded individuals to connect, but provides relatively safe environments where expressing personal beliefs is nurtured rather than ridiculed, and the option of anonymity is available but rarely opted for. Virtual fraternal forums afford validation and support.

While researching for my first book, finding legitimate information about UFOs and encounters was equal to a scavenger hunt. Today, there is more dependable data at our fingertips than anyone could ever digest.

This enormous change has occurred in less than fifteen years. It is inspiring, and stirs the imagination as to what else awaits us in yet another fifteen years.

The Term "Close Encounters" - I would be remiss if I did not address the term "close encounters."

The expression "close encounters" was not coined and claimed by Ufology to specifically represent contact with UFOs and alien beings until the early 1970s by UFO researcher J. Allen Hynek. A few years later the phrase solidified into modern nomenclature by the enormous success of the movie "Close Encounters Of The Third Kind."

Unfortunately, the phrase has been denied its full potential, and by that I mean: We are not even close to understanding the true nature of the numerous beings we interact with on and off world, on a daily, and more often nightly basis. It is my opinion that the experiences I have had and the entities I have personally encountered and found in my research, cannot possibly fit into one specific category and we are already over our heads with terms that are more confusing than they are helpful.

Categories such as: According to Wikipedia 2010 we are now up to 9 versions of CEs (close encounters). We have terms for entities infinitum; alien, ET, UFO occupant, extraterrestrial, interdimensional, extra-dimensional, multidimensional, ultradimensional, hybrid, circle maker... watcher, founder, ancient, and on and on. To further baffle, some researchers argue aliens have nothing to do with UFOs and UFOs have nothing to do with aliens. Most encountered beings defy our physics, move through walls, and appear and disappear in thin air.

Depending on our spiritual make-up, education, and beliefs at the moment of contact, an entity is labeled; ghost, angel, fallen angel, demon or devil. They are our higher-selves or lower-selves. They are illusions, or are they delusions?

New agers describe them as; past selves, future selves, parallel selves and counterparts... astral travelers, doppelgangers, spirits or spirit guides... shape-shifters, gods and goddesses or Jesus.

Conspiracists claim they are deep black military robotic constructions, mind control holographic/virtual reality projections, psychic and/or drug induced hallucinations.

So where do they come from?... inner space or outer space, inner Earth or probable Earths... far reaching galaxies, or parallel dimensions, alternate timelines or heaven or hell... Zeta Reticulum or Kansas?

The only true answer is we *really* don't know who or what they are or where they come from. To try and force them into a box to mollify our impotent understanding is a disservice to our evolutionary path. I vaguely recall a Buddhist practice of taking a walk in nature and absorbing the experience without labeling anything observed.

Which brings me to the last chapter of this book which is of wider scope. It contains all the stories that do not fit the narrowly defined profile by J. Allen Hynek, but they are indeed "close encounters" of curious kinds.

The Evolution Of Encounters - Once again I have to say, the evolution I see taking place in the spirit of humanity, in relationship to encounters, is as intriguing and important as the contacts themselves.

There is excitement as more and more people discard fear and allow themselves to inwardly believe and outwardly express realities we have secretly known to be true for so long.

There is power in numbers, and the numbers are growing... if by chance or if by design.

Valid information is flowing in from every direction... we are banning together to educate ourselves and each other and to reveal essential truths. We are openly battling ignorance and suppression... and it is working.

My experience of our collective state of being at present, is that we are participating in an accelerated heightening of consciousness on this planet. Time is compressing, conversations are maturing, values are being reassessed, and goals redefined. Greater energy is being focused simultaneously on larger universal concerns as well as deep personal growth.

It feels as though we are in the nexus of a long awaited birth into a great new sphere of potentiality... one that is anxiously overdue. It is on the forefront of the minds and lips of anyone paying attention. I contend we are romancing the concept of a quantum leap, flirting with consciously participating in the grander multidimensional cosmos.

Need we be reminded we must sever the remaining tethers to fear, greed, selfishness and separation in order to play at such a level.

I choose to believe we are succeeding in this radical change. However challenging, the ethereal bleachers are loaded and the chorus of the angels can be heard with the heart.

We are anything but alone. Move in any direction and you will find a fellow traveler... left to right, up/down, in or out. We travel in different modes, in space and in time, in body and in mind... and in spirit. We move from moment to moment, from desire to desire, from love to love, from lifetime to lifetime.

Sometimes we are the visitors, sometimes the visited, and sometimes we bump in the night. At times our crossings are as delicate and brief as a breath across your cheek, but the impact can be profound and everlasting.

The birds migrate the ancient energy streams of the seasons. The planets embrace their familiar orbits, but you and I?... we journey the way of the great mystery, and all is as it should be.

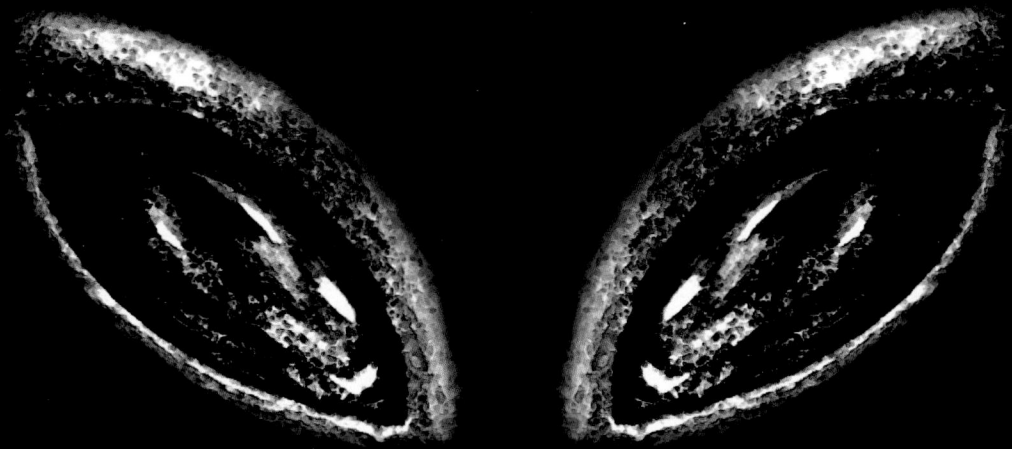

The Art Of Close Encounters

Forensic Encounter Art

Christine "KESARA" Dennett - was born in Salt Lake City, Utah, September 26, 1955. She grew up in an artistic atmosphere, always encouraged to create art. In 1972 She was given the Sanskrit name "Kesara" by a Buddhist Monk, Thien Thic An. It means "beautiful flower."

Since 1986 Kesara has been a UFO illustrator for experiencers and major investigators. Her renderings have been published in books, magazines, on the internet, and featured on several major national television networks. Among the experiencers and prestigious investigators that have elicited her talent to express their fascinating accounts and real facts about UFOs and Extraterrestrial are; Preston Dennett, Linda Hamilton, Timothy Willie, Phillip Krapf, and many more. She assisted three time Emmy award winner, Vince Depercio, investigating UFOs and contacting extraterrestrials for documentary work.

Kesara is not only an objective observer illustrating for others as a forensics sketch artist, but she has had her own personal experiences as well. As a result of her work in the UFO field her encounters have multiplied. Kesara has produced over 200 stunning illustrations based on actual experiences reported and investigated. Kesara's primary commitment to her clients is to produce images that are as emotionally communicative as they are physically accurate.

"Anything is possible in this Universe, we must strive to open ourselves to the infinite realm of existence." - Kesara.

Eliad - The Eliad are a race of extraterrestrials I have illustrated for three clients. These beings have been witnessed on ships as well as in dreams by the experiencers. Tall and elegant (about seven to eight feet), they wear Egyptian style clothing and elongated helmets to protect the engorged, fleshy back side of their skulls. They are gentle creatures and their movements are graceful and ethereal. Working with humans on Earth, they research not only the physical and environmental aspects of human life, but explore our intrinsic natures as well.

Eliad

Kesara

Zetas - Zeta is one of the names that has been
adopted for the gray type extraterrestrials because
they have told many abductees or experiencers
that they come from the binary star system, Zeta
Reticulum.

In my twenty five years of research, the Zeta has been
described as follows: clinical, cold, mechanical, and
loving. I have received countless reports from
people who claim their relationships with the grays
encompass a range of emotions from pure terror to
everlasting love.

I personally witnessed them moving in and out of our
dimension out in the mountains near the 210 Freeway
in Southern California, although I do not recall being
taken aboard a ship.

Zetas

Kesara

23

Hybrid & Family - The subject of this illustration is the hybrid. Many species of extraterrestrials are known to be breeding with humans. The grays (or Zetas) are the most commonly reported race conducting these types of programs.

According to some of the experiencers I have interviewed, it is believed the grays are creating genetically engineered, hybrid races, made-up of characteristics of the grays and characteristics from all Earth races, but may also include other ET species as well.

Therefore it is easy to understand that the hybrids appearances are varied, unique and interesting.

Two things I found most fascinating in the reports about the hybrids were; their powerful telepathic abilities, and the incredible amount of love that emanates from them. In many cases they have met their Earth parents and expressed gratitude to them for their participation in the breeding program.

Hybrid & Family

Kesara

Mantis - This mantis was commissioned by Preston Dennett. It was witnessed individually by members of a group of people who were independently under-going regression sessions with the same hypnotist.

They eventually got together to share notes and discovered that they all had similar experiences. A book was written about one of the group members entitled "The Coronado Island Incident" by Michael J. Evans and Preston Dennett.

The description of the mantis extraterrestrial was very precise in that its arms were disproportional in length to its body, and the eyes were extremely large exactly like the insect mantis on Earth. Its skin is notably pale and it stands to be at least seven to eight feet tall.

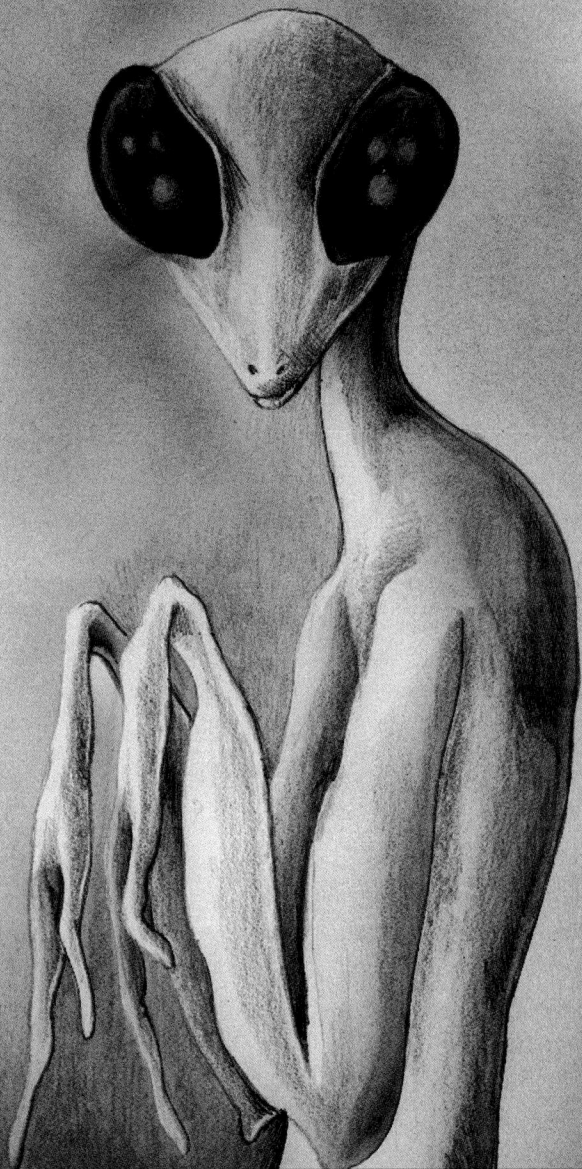

Mantis   Kesara

Praying Mantis w/ Triangle Pin - Praying mantis extraterrestrials are very tall insect like beings similar to the praying mantis insects on Earth.

They work in tandem with other extraterrestrials, such as the gray type ETs. This is one of my first drawings from the praying mantis series I have illustrated.

The mantis has been a source of fear for many abductees/experiencers, and a source of fascination for investigators.

They typically have a cold, clinical, detached attitude which makes them perfectly suited for the obvious scientific research they are conducting on their human detainees.

Mantis w/ Triangle Pin

Kesara

Oldest - In Simi Valley, California this older gray extraterrestrial visited a young gentleman every night for two years. The experience was very intense for the young man.

There was a great deal of physical research performed on his body. He sustained some injury which indicates these beings lack a degree of sensitivity towards their subject's bodies and pain thresholds. Despite the negative aspects of his visitations, the young man doesn't regard this particular being with fear or resentment, but surprisingly described him with a kind of affection.

When I went to visit with the client his energy was very... expanded, intense and infectious. I experience this abounding energy with experiencers all the time. Being in their presence is sometimes equivalent to drinking two double espressos!

Oldest   Kesara

Green Reptilian - My portfolio is full of illustrations produced from witness descriptions of reptilians.

This particular art piece was my first portrayal. I have received at least twenty different descriptions of the reptilians, with just as many diverse elaborations about their demeanors and the types of relationships the experiencers have with them. The amount of information on them is quite extensive.

Repeated over and over again, is that they are not from outer space, but rather... they are from inner Earth.

Green Reptilian

Kesara

Nordic Female - The most common description
of these amazing beings is that they resemble the
Nordic folk here on Earth.

They are anatomically ideal, beautiful, tall and
muscular... the personification of perfect super-
humans. Some of my clients are so protective of the
Nordics that I have signed privacy contracts before
working with them.

There is a strong suggestion by many that our DNA
has been manufactured by them.

Nordic Female

Kesara

Nordic Male - I've had a very interesting connection of my own with one of these humanoid ETs. He seems to come around when I work with clients that have more mystical types of experiences with his kind.

And to be quite honest, I too feel the compulsion to keep our relationship personal until the time is right when I can share with the public his purpose for working with me.

Nordic Male

Kesara

Galactic Family - Represented here are a few of the most commonly reported types of UFO occupants, and perhaps a few less common types. I have included a reptilian, a praying mantis type, a Nordic female, a human-like male (possibly a hybrid), a stocky blue being, short and tall beings called grays, an Oriental-looking hybrid type, and a small greenish being with features of grays and reptilians.

Encounters with UFO occupants, or generally humanoid beings that seem to be associated with UFOs, are a central part of the UFO mystery. For simplicity I refer to these beings as "aliens." I know that some readers may object to this term and I don't mean to distance or alienate ourselves from the beings, but "extraterrestrial" usually implies an origin in other star systems when it may be more plausible that some of these beings come from the future or from parallel quantum realities here on Earth.

For the particular species or group of species most often encountered, the term "gray" is the one that has stuck. In truth some of these beings are chalky beige, white, tan, blue-gray, or grayish green.

We could also call them the gray people, and the reptilians the lizard people, to recognize them as non-human peoples. Nevertheless, "gray" brings to mind the large, black eyes, bulbous heads and small bodies that characterize all "grays", so I feel the term is appropriate for our present level of knowledge about these creatures. "Zeta Reticuli" commits us to a particular theory of their origin that cannot yet be verified. Until a better term is found I will call them "grays."

I hope that other artists with an interest in doing this kind of research will find this image useful. There is much interesting work awaiting, including interviewing additional witnesses and further examination of the currently available information.

As artist-researchers, our goal should be nothing less than accurately documenting the appearance of each type of alien that has been consistently described. This begins with giving deference to individual eye-witnesses and attempting to visually portray, as accurately as possible, what they remember seeing.

I also believe in making our findings widely available to the public, ASAP, as realistic images of aliens have the potential to generate great interest in the alien presence. This, in turn, may help to encourage intelligent discussion of the phenomenon by the general public, which may be a prerequisite to the phenomenon receiving the serious scientific attention that it demands.

Reptilian Types

Drac - B.H. says the claws are ≥ 1 inch long ~1.5 inches. B.H. did not notice any webbing. the Dracs do not smile.

Drac resembles "Predator" from the Schwarzenegger movie - but with a reptilian head.

Drac
Draco
Draconian

Composite Description:
Large eyes with catlike pupils.
Protruding lizardlike snout.
Scaley, greenish skin.
Muscular body. ~ 6' tall
4 fingered hands - or 3 fingers
and 1 thumb with dark claws
~1.5" long. Possibly some
webbing between the fingers.
Narrow waist in proportion
to chest. Mechanical movement.
Aggressive behavior.
Long arms. Reaches out.
Eyes almond shaped &
slightly slanted.

- How large is the cranium?
- What does the neck look like?
  skin may be lizardlike
  or crocodile-like.
- What is the appearance
  of the brow areas,
  fore head, scalation &
  skin fold patterns?

In addition to the "Draco soldier" types are there reptilian types which more closely resemble the little Greys? skin is wrinkled around the eyes

vectra? draco?
Bent forward posture
Narrow waist
Thick Thoracic cavity & Thighs
Powerful Limbs
wide Shoulders
Beaklike snout with fleshy appendage dangling from upper jaw
muddy green, dry, scales.
Birdlike or Lizardlike
Talons.
Head a little larger
is the back than
a human head.

Eyes are large, rounded, oval, or almond-shaped, w/ slit or starburst pupils and yellow irises.

Hands w/ 3 or 4 fingers and an opposable thumb.

J.S.C. Says this hand is accurate - the webbing may be even more obvious.

Based on face on p. 96-B of Matrix II, somewhat resembles a lizard embryos has that neotenized look.

This mouth looks too "smiley"

Linda Moulton Howe says the reptilian types are often 6 and 7 feet tall.

John S. Carpenter says the neck may not be as thick as shown here and they definitely don't Smile. The mouth is like a lizard or snake. He also says that some have a ridge down the forehead.

Source: DQ. 7-1-94
Feet: Has humanoid feet w/ ankle knobs heel, arch & ball to the foot but foot is stockier than human in keeping w/ large mass & height.
~ 4 toes, no little toe
Some webbing - Toes not small or well defined. Not distinctly pronounced - segmented like human.
"Clublike" Feet
Toes.
Claw appears larger than normal for this foot.

Kraton has a serious look although protruding snout does give mouth a curve - a perpetual smile.

Eyes not so wide open - more of a squint - with lid partially closed over eye.

Larger on head & Face than shown.

Protruding "v" Shape to powerful lower jaw.

Kraton
Powerful jaw muscles
Head res embles a Teenage Mutant Ninja Turtle.
Yellowish, greenish skin?

Kraton's mouth does curve like a smile because of protrusion of muzzle.

Pointed a little bit higher & more pointed than originally drawn - with a barely visible ridge or head-crest

Nostrils small, as shown, but slightly lower - slightly closer to tip of the snout

J.S. Carpenter says this face is fairly accurate and that it is one case the Profter. Snout may have protruded even further. Cross-hatching is a good technique for representing the scales.

John E. Mack says that by and large the reptilian types seem very mechanical - robot-like - and unemotional. They may be more muscular than the Greys.

Based on p. 96-B of Matrix II by Valdamar Valerian

Large scales on neck & face.

Based on Face shown: UFO Coverup Live
- Reptilian Features added to interpretation of Barney & Betty Hills alien.
- Profile resembles Vectra.

Draco "Soldier"

Kraton

Reptilian Types

David W. Chace

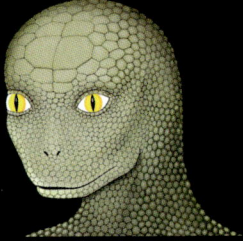

Renjeck - A highly intelligent, male, reptilian being. His body is covered with a thick hide of small, bead-like scales of dark greenish-brown and lighter olive tones. His hands have three fingers and a thumb, like a human hand minus the pinky (but with claws and beaded scales). He has a short tail, about two feet in length.

Unlike many reported species of reptilians he does not have a ridge on his forehead, but rather a patch of large leathery scales or plates. There is a series of small bumps that runs down his spine and tail.

He sometimes wears a black cape or other clothing for disguise. He has a good deal of manual dexterity and sometimes carries a backpack with an attached gun-like rod that is able to induce paralysis. The back-pack apparently contains the power source for this device. At times, he seems completely merciless and amoral, but on rare occasions he has also shown himself capable of expressing deep sympathy and remorse.

He is highly skilled in the use of telepathy and telekinesis. He can transform himself into a red ball of light.

His job is complex, but in short, he is an agent for the grays involved in a global (Earth) policing operation aimed at removing intrusive alien species.

He recruited and trained Kenny as a child in order to help him in his duties. In several instances this involved Kenny being used as a decoy or "live bait" in what amounted to sting operations.

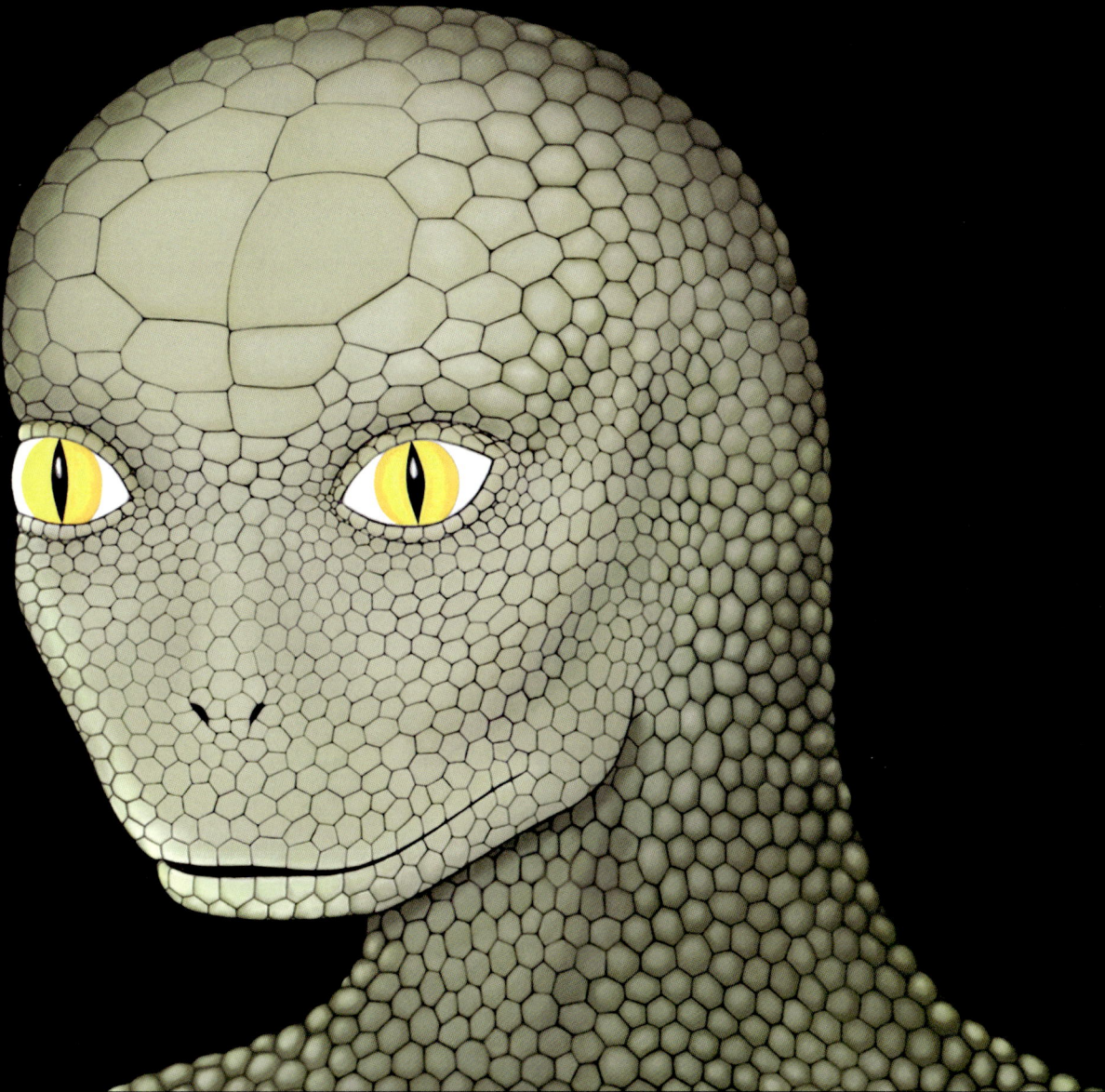

Renjeck

David W Chace

43

Reptilian 1 - This male reptilian seems to have the ability to shape-shift into a human form.

When he first appeared to Dana (pseudonym) he had the appearance of a handsome, blond man. In this form he even seemed human to the touch, possibly suggesting a physical transformation, rather than mere visual trickery.

In an instant, however, he returned to his reptilian form, which was a shock, though the loving telepathic communication from him helped Dana to overcome her fear.

When he left Dana, he disappeared as if being sucked back into another dimension. Perhaps this is a form of teleportation. He seems to be just at home in the material plane, as he is in the astral realms where some of Dana's contact with him took place.

His skin was smooth, soft and firm, like the skin of a snake. The color was a mix of greens, browns and

grays. The ridges above his eyes and on his forehead were a lighter color than the surrounding scales. There were also ridges of lighter color below his eyes, but these were less prominent.

The mouth was a wide slit. It opened a little when he grinned at Dana, and she could see that he had small, jagged, uneven teeth.

Something about the appearance of the lizard species called the "bearded dragon" reminds Dana of a miniature version of him.

The tendons on his neck were very pronounced. There was no sclera (white) visible in the eyes, only iris and pupil. The iris seemed to have the ability to change color, perhaps due to the presence of chromatophores responding to changes in mood or arousal.

His interest in Dana may relate to past-life connections, and possibly to an intention to create a reptilian-human hybrid.

Reptilian 1   David W. Chace

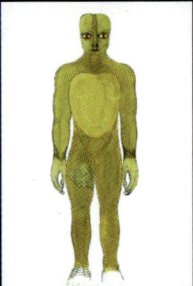

Reptilian Body - Charlene Smith (pseudonym) has been abducted by more than one group of aliens, including a reptilian species.

This individual is the highest-ranking reptilian she has seen. Charlene regards him as a fleet commander. He is definitely an alpha male. He has militaristic attitude.

He is highly disciplined and self-righteous. He regards his race as superior to human beings and thinks of humans as mere animals, the mutts of the universe. He is an advanced reptilian-human hybrid, the result of past transgenics experiments involving humans and at least one ancestral reptoid species. He can transform himself into a white ball of light.

He is about six feet tall with a lean, muscular body covered by soft, smooth overlapping scales that point downward. The texture and feel is similar to snakeskin. The skin color is a neutral khaki (mixed tans, greens and grays) with a yellowish cast. He has a hard, smooth breastplate or plastron that covers his chest. This resembles a turtle shell, with several, large plate-like scales, but it becomes somewhat softer and more flexible lower on the torso. On the lower abdomen it grades into the surrounding skin.

His head has a central ridge that is darker in color than the surrounding skin (the opposite of what you find with some other reptilians). His ears are curved slits about 1.5" to 2" long. He has some kind of vestigial slit between the nostril and eye on each side of his face, perhaps like the heat-sensing pit of a viper.

His eyes are large with a bright golden yellow iris. They are not slanted, nor deep-set. There is only a subtle indentation demarking the upper eyelid. There are four thick cords or tendons that run down his neck. His hands have three fingers and a thumb, with claws.

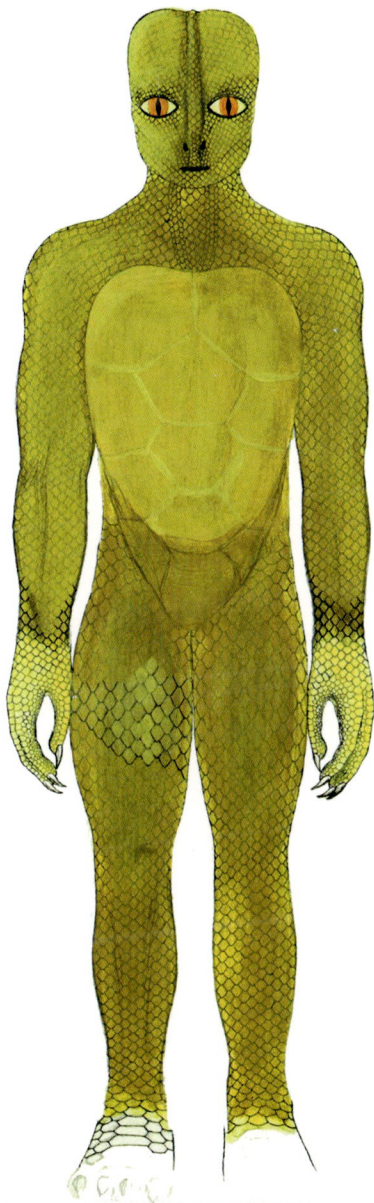

Reptilian Body

David W Chace

Reptilian Gray - I call this general class of being a reptilian-gray.

This type of being appears to be a hybrid between the scaly skinned reptilian beings (that typically have a more human-like build) and the small, big-headed grays.

Several types of beings with this combination of features have been reported over the years. Some have obvious scales and some do not. The vertical slit pupil does distinguish them from the "regular" grays, which generally have a solid black eye without an obvious pupil. They may operate as their own group, or they may serve as workers for other types of beings, typically under the supervision of some taller type of alien.

The being in this image wears a sash and a shoulder patch with a symbol or insignia. It may be a kind of military uniform.

Grays and reptilians have both been perceived as organized and militaristic, so it would make sense that a reptilian-gray might also have such a behavioral disposition.

Reptilian Gray   David W Chace

49

Communion - This type of being is a form of tall gray, although in this case it is more of a yellow-beige color.

It was made famous in 1987 with the publication of Whitley Strieber's best seller "Communion," which featured a painting of a female being of this type on its cover.

Typically, the cranium appears somewhat larger than was shown on the cover of "Communion" and I tried to represent that with this image.

Also, these beings have hardly any lips and only a very slight nose. This kind of tall gray has a more elongated appearance than the majority of reported grays.

They also have a more highly developed mind and a richer capacity for emotion. Tall grays typically are in charge when seen with a group of grays, and they may serve as teachers, doctors or physicians, rather than just workers or gofers.

Though these beings can be male or female, there are no visible differences between the sexes, and gender is usually just an impression the abductee gets from telepathic communication or from a certain feeling the being emanates.

Reports suggest that at least some beings of this type levitate rather than walk. The means of levitation is unknown, and could be either techno-logical or paranormal.

I should also point out that the tall grays are quite distinct from praying mantis type (which are much more insect-like in appearance) or reptilian aliens (which have scales and distinctive vertical pupils), although some writers have confused them in the past.

Old Gray - This old male gray was the leader. He was the familiar alien who was in charge of Martina (pseudonym) during her involvement in the breeding program that the grays are conducting.

He is taller than the worker grays but shorter than the average human being. His body is slender. His eyes are black. His skin is gray-beige and darker on the face than the cranium. His face has numerous small pockmarks, like subtle indentations in clay. He has just a little flesh over his bones.

Some people believe that the grays are merely robots, but this being is a biological entity, not just a machine, and the marks on his face may be a sign of his advanced age.

An interesting anecdote about this being: Martina hasn't seen him in years because her involvement in the breeding program seems to have ended. However, a young friend of Martina, Donna (pseudonym), who *is* currently involved with the grays, saw him in Martina's home while she was visiting.

Donna wanted to go tell Martina that her alien friend was there, and she should come take a look but the old gray telepathically told Donna (paraphrasing) that he doesn't want Martina to see him anymore because the time for that has passed.

Apparently the old gray would rather not disrupt Martina's present life with another sighting and any difficult memories it might bring back.

Old Gray   David W Chace

Military Hybrid - This male being was seen during an abduction that took place in 1995. The location was a military base somewhere in the American Southwest. He wore a uniform that reminded the witness of a Confederate uniform from the Civil War.

The uniform had a mandarin collar and a lapel, and was green like split pea soup, but with a hint of gray. There might have been additional detail like piping or cording along the edges, and perhaps shiny buttons or an emblem of some sort. His skin was ash white, like porcelain. His neck was thinner than white, like porcelain. His neck was thinner than normal by human standards. His lips were quite thin and his ears were very small. There were subtle age lines around his eyes, and his irises were a crystal gray color. Based on his appearance, he might be a hybrid, or a transgenic mix of human and Gray characteristics.

His personality came across as cold and emotionless. This being was acting as a doctor/supervisor, but oddly, he seemed to be in charge of human military personnel.

Yurani

David W Chace

Estartleah - This female being emanates a feeling tone of unconditional love. Her body glows from within with a bright white light. She seems to be a solid, physical being, but also appears to be able to de-materialize and take her physical body with her into the astral realms. She may be capable of invisibility.

She levitates several inches above the ground and moves by floating from place to place.

She is very slender and has a delicate, elongated appearance. She wears a long, white gown or robe made of a light, gauze-like fabric. Her hair is long and wavy, but has a fleecy or wooly appearance, and is very light blond, almost white in color. It looks like an Egyptian style wig, and Marlo (pseudonym) suspects that it may be a wig.

She has vertical, slit pupils that are blue in color, and blue veins visible in her temples. Marlo thinks of her as more feline than reptilian, but the slit pupils do

give one pause. She also has a nictitating membrane that slides up over the eyeball when she blinks.

Her arms and fingers are very long and thin, and her four fingertips are rounded and slightly flattened, with gripping pads, kind of like a tree frog. She has a pointed chin and a large cranium. She has small teeth like new bud corn kernels.

She has been Marlo's primary contact since childhood, and Marlo thinks of her as her "space mother." This glowing female being seems to belong to a race that has ascended to another plane of existence, yet some of them do reside in a vast space colony to which Marlo has been taken. Also, when they show up physically in our world, they are vulnerable.

The female is generally accompanied by a small male being who also glows, but not as brightly. He is her bodyguard and traveling companion.

Estartleah   David W Chace

59

Blue - I painted this image based on a handful of reports of stocky blue-skinned beings. These beings tend to be short, but the height varies from about two feet to a little over four feet, depending on which type is being reported, as it seems that there are a number of similar-looking, possibly related types.

Their squat bodies seem to have more skin than they should need, and on the face this extra skin is arranged into several horizontal folds and furrows.

They have bulging black eyes with heavy lids that blink often. The nose is wide with wide nostrils, and sometimes described as squashed-looking. The lips are wide and thick.

These blue beings certainly don't look anything like your typical, slender, thin-lipped grays. In some instances the blue beings seem to move by levitation, in others they walk with a funny waddle.

Passing through solid objects appears to be child's play to them. They are most often seen wearing dark, hooded monk's robes or jumpsuits. The hands typically have three or four stubby fingers, perhaps with fewer knuckle joints than human fingers.

The skin color has been described as ranging from primary cobalt blue, to dark navy blue, to shiny black, depending again on which type is seen.

Their role in the abduction phenomenon usually seems to be as workers or helpers, but such beings have also played a role as spiritual teachers in a few reports. They are sometimes encountered in the presence of grays, and sometimes in other contexts.

More people have seen the blue beings than you might think, but as far as I know, no one wrote about them prior to the release of Whitley Strieber's best seller "Communion."

Blue   *David W Chace*

Mantis - This being is one of several reported types of aliens that resemble the terrestrial praying mantis. It has a very bony appearance, and may possess both an endoskeleton and some form of exoskeleton.

Beings such as this sometimes serve as doctors. They are definitely above the small gray beings in terms of authority, and their minds are very powerful.

Mantis   David W Chace

Glowing Goblin - One of the strange, glowing, goblin-like creatures encountered on a farm near Hopkinsville, Kentucky, on an August night in 1955 after their craft landed in a nearby field.

These creatures were about three and a half feet tall. They were bipedal, but would drop to all fours when they ran, as the arms appeared to be longer than the legs.

They were good climbers, and bullets did not seem to injure them. These beings had large ears, wide mouths, wide-set eyes and clawed hands. No scales were reported, so one couldn't say they were explicitly reptilian. No hair was reported, so one couldn't really call them anthropoid. They didn't resemble grays or insectoid aliens. They didn't look remotely human.

Whatever they were, such goblin-like creatures are not common among reports of alien contact. Certainly, other cases of glowing beings have been reported in more recent times.

Is this unexplained glow a form of bioluminescence, something technologically generated, or something paranormal? Perhaps the glowing being is in some kind of "energized" physical state in which light flows through it into our reality from some higher dimension or other plane of existence.

It seems clear that, at present, we don't understand what this process actually entails. We also don't understand what role anomalous types of aliens, such as this, might play in the larger UFO phenomenon.

Glowing Goblin   David W Chace

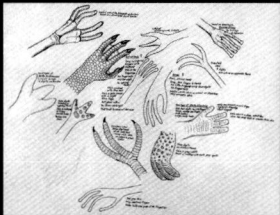

Alien Hands - Alien beings have several different types of hands with different numbers of fingers in different arrangements.

Reptilians typically have a thumb and three or four other fingers arranged as a human hand, but they have scales, thicker nails or claws on their fingertips, and sometimes some webbing between the fingers. Insectoids, such as the various praying mantis type aliens, have the most alien-looking hands, and perhaps the greatest variety of different kinds of hands of any general category of aliens.

Grays, in most cases, have slim, dexterous hands, sometimes with small claws and sometimes without. Some types of aliens, including some grays, have stubby or mitten-like hands.

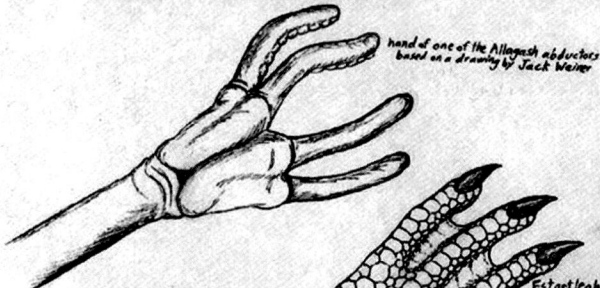

hand of one of the Allagash abductors
based on a drawing by Jack Weiner

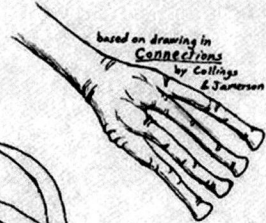

~Robot grasping hand, 3 digits

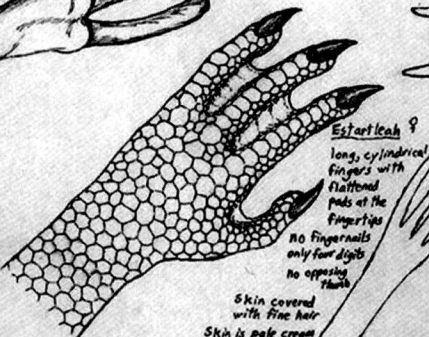

based on drawing in
Connections
by Collings & Jamerson

Estartleah ♀
long, cylindrical fingers with flattened pads at the fingertips
no fingernails
only four digits
no opposing thumb

insectoid claw
3 digits
one acts as an opposable thumb

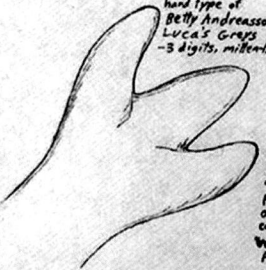

hand type of
Betty Andreasson Luca's Greys
-3 digits, mitten-like

Skin covered with fine hair
Skin is pale cream with a slight blue tinge
but glows with a brilliant white light
that tends to wash out her color

Yerani ♀
Frail, skinny hands
long, thin fingers & thumb
no little fingers—only four digits
no fingernails
pasty white skin w/ a hint of flesh tone
very smooth skin

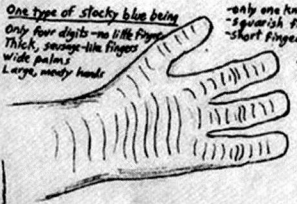

Three digits included an opposing thumb
Palm & underside of fingers contain fuzzy Velcro-like pads

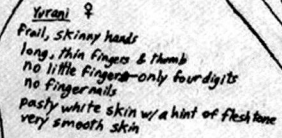

One type of stocky blue being
Only four digits—no little finger
Thick, sausage-like fingers
wide palms
Large, meaty hands

-only one knuckle in each finger
-squarish fingertips
-short fingers

Skin color is a deep, cobalt blue
Skin has many folds or creases—thick skin

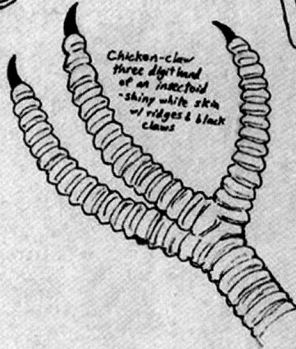

Chicken-claw three digit hand of an insectoid
-shiny white skin w/ ridges & black claws

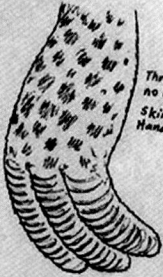

Three digits
no distinct thumb
Skin is grey-white
Hand is mottled with dark grey spots

Pale grey skin
long, dextrous fingers
fleshy bulb-like pads at the fingertips

Pretty Feet - I was visiting someone who has had numerous experiences. She took me to her bedroom and showed me three prints on her closet door mirror.

There were two obvious hands and an odd looking one in between them. I asked her what that was, and in an odd voice (as if it wasn't her talking), and with no emotion, she said, 'That's a foot'. So this drawing is based on those prints.

The knuckle bones appeared to be more prominent than ours, and the fingers much longer. The thumb pads were cocked in at an angle that we couldn't achieve.

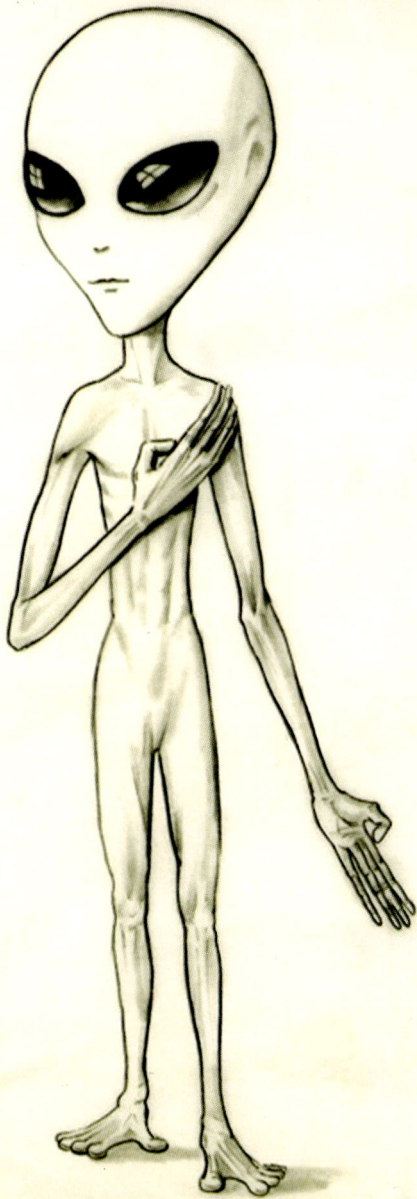

Pretty Feet   Corey Wolfe

Gray Salute - This was done, just to see how my other sketch would look in color. Brad Steiger bought the rights and put it on one of his books entitled **The Other**. It's painted with an airbrush (way before the days of computers).

Author's note: Even grays look better in color!

The Phoenix Lights

Larry Lowe

75

Paradigm Shift - My story is made up of two emails to Kim, the author, on Facebook. She challenged me to allow her to publish them as my story. I agreed.

First email to Kim: My paradigm has changed at least three times in my life. The first time was on my birthday, May 6th, 0300 to 0400 hours, in the mountain military reservation belonging to Kirtland Air Force Base, Albuquerque, New Mexico. Secret clearance required, we did war games and I wandered out of the designated areas.

What I saw with night vision could not be explained by normal physics. As a GySgt in a surface to air missile battalion (expert in military aircraft I.D.) and part time flight instructor for small, single engine airplanes, I experienced what could only be explained by off-world technology! I witnessed gliding "machines" changing shape and form in between mountains of solid rock in the cold night air. There was no sound, just elegant dancing in the sky and the base below. It was then I realized - I know nothing about the world we live in. I knew nothing... and felt so small.

Second email in response to Kim asking me to use the previous email in the book: Oh geesh, everyone has a story, I read your book, "Beyond My Wildest Dreams: diary of a UFO abductee." My story is lame compared to your experiences; but like I have said to others, I'd be happier knowing the truth of the world, no matter how terrible or different from my paradigm. I choose to know the truth verses living a life of blissful ignorance, consumerism and pleasure... no matter what the outcome. I have mentioned this event only once in an English 102 college paper to a cultural anthropology professor. *(Kim offered to try and produce a rendering for me in Photoshop).* Yes, I'm a Photoshop geek too, but how do you display a machine which appears as graceful as a life form, knowing it is technology?

It's like they harnessed the power of life and integrated it with their machines! Here you have a seasoned Marine, an expert in military aircraft identification, part time flight instructor giving classes in airplane flight physics, who then sees an alien world between the rocks of two tall mountains in the dead cold of night where sound waves carry every whisper. Nothing was audible but the pounding of my heart and breathing, the airstrip and base below dancing with magical forms that changed shape and glided through the air effortlessly. It was obvious they used none of our laws of lift, drag and thrust.

I feel like an insect, an ant to the superior human form. I was not supposed to be there... an outlier... like I don't matter in the schema of life. We are living in a bubble and I am tearful as I write this to you... a confidant, someone who may understand my tears as they stream from my cheeks.

This life is just a game and there are other forces at work... we know nothing. We are so insignificant. I'm sorry, the tears take me away. It's wonderful to know another person who may understand me. How do you cope? Monday night football? Learning a new language? Just how long does it take to accept the true paradigm and to realize how we fit in the schema of life?

It would be so much easier learning this as a child. Thank you for listening my virtual friend! I still say "give me the truth over bliss", I demand it! I need to know I am not insane. I feel two feelings now, violated and relieved. How those two feelings go together, I don't know. I should go.

Paradigm Shift   Dalton Bagby

Seeing The Light - My father, grandfather and I witnessed this sighting at my grandparent's home in Rome, Georgia, US in the mid 1970s.

We were all sitting in the living room talking, watching TV, etc., when we were immediately tuned into a sound from outside, a sound we knew very well, the thumping noise made by the rotors of a helicopter.

My dad and granddad took me outside to see the helicopter where we made our way into the front yard for better viewing. The sky was cloudless and the view unlimited, many stars could be seen against the clear night sky.

Upon first glance, I noticed there not to be one, but many helicopters flying around in a circular flight pattern. No markings were visible on the aircraft. I was very surprised and extremely excited realizing there was not just one, but five helicopters.

My granddad asked, "What in the world is that?" as my dad and I also caught a glimpse of "a light!" It was about as large as a baseball held at arms length, and moving very slowly and totally silent.

The round ball of light would flicker and pulsate like a strobe light, then completely disappear only to re-appear in another section of the sky. The object would constantly change color from a blue to a white, and it would hover in one place, move at a snails pace and then "jump" from one section of the sky to another.

All the while, the helicopters were trying to keep pace with it. It was playing a game of "cat and mouse"... it was defiantly responding and interacting intelligently with every move or maneuver that each helicopter made or attempted!

We watched for about ten minutes, as the object "toyed" with the obvious mismatched and ill equipped helicopters. The technology and ability of that object was incredible.

As we were continuing our jaw-dropping aerial encounter, the object performed another maneuver that was just as incredible as all of the rest... the object went from its standing position in the sky to almost an instantaneously rate of blinding speed, then vanished out of sight!

Gray Salutè   Corey Wolfe

Sightings Encounter Art

The Phoenix Lights - On March 18, 2007, the Prescott Daily Courier printed a special story written by investigative journalist Leslie Kean in which former Arizona Governor Fife Symington claimed to have witnessed an "enormous and inexplicable" object in the evening sky of March 13, 1997 - the night of the Phoenix Lights.

In less than a week Symington was interviewed by CNN and Phoenix television reporter Scott Davis at the site of his observation, a small park to the west of Squaw Peak in north central Phoenix.

In November of 2007, a brief description of the sighting appeared at CNN.com under Symington's by-line which is essentially the same as his remarks at a conference in Washington DC organized by Kean and filmmaker James Fox, which the Governor moderated.

The Governor's account has remained sparse and consistent since he first admitted having seen the craft. He says in part: "In 1997, during my second term as governor of Arizona, I saw something that defied logic and challenged my reality.

I witnessed a massive delta-shaped, craft silently navigate over Squaw Peak, a mountain range in Phoenix, Arizona. It was truly breathtaking. I was absolutely stunned because I was turning to the west looking for the distant Phoenix Lights. To my astonishment this apparition appeared; this dramatically large, very distinctive leading edge with some enormous lights was traveling through the Arizona sky. As a pilot and a former

Air Force Officer, I can definitively say that this craft did not resemble any man-made object I'd ever seen. And it was certainly not high-altitude flares because flares don't fly in formation. What I saw in the Arizona sky goes beyond conventional explanations. When it comes to events of this nature that are still completely unsolved, we deserve more openness in government, especially our own."

In an interview with CNN, Symington described the craft as 'otherworldly'. As part of a project to document the sightings that night, I contacted Scott Davis. I had, from previous reports, developed a path for an enormous triangular object.

Using Google Earth, I put the camera at the location in the park where the interview was filmed and pointed it at Squaw Peak. I then positioned a model of the craft scaled to a mile along the trailing edge of the delta in a position that matched other witnesses sighting's reports.

The result is that the gigantic triangular craft, as distinct from the V-shaped wing reported by Tim Ley and others, would have indeed been visible to anyone in the park on that evening.

The lighting configuration is as seen in the only known video of the large objects the overflew the Phoenix metropolitan area between 8 and 8:30 p.m. on the night of March 13, 1997. It is also derived from a drawing of an observation by Sue Watson presented on the National UFO Reporting Centre website.

This attempt at a forensic recreation of Gov. Symington's sighting indicates that if he was where he said he was, looking in the direction he said he was, he would have seen something that matches descriptions by other observers that night.

Seeing The Light   Art by Kim Carlsberg   David Patrick Kuhlman

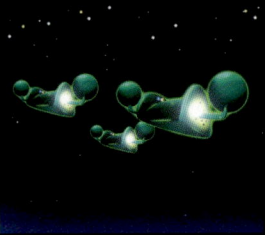

Emerald Trio - La Frette-sur-Seine (15 miles from PARIS) 7th, July 2004, 4:05 am.

Testimony (Witness): Daniel C., 50 years, professional musician. These objects passed just over the testimony (witness) when he was driving on the highway to his home from Paris.

Investigators: Georges Metz and Jean-Claude Venturini. Published in the magazine "Lumières Dans La Nuit" Nr 377 (Feb. 2005).

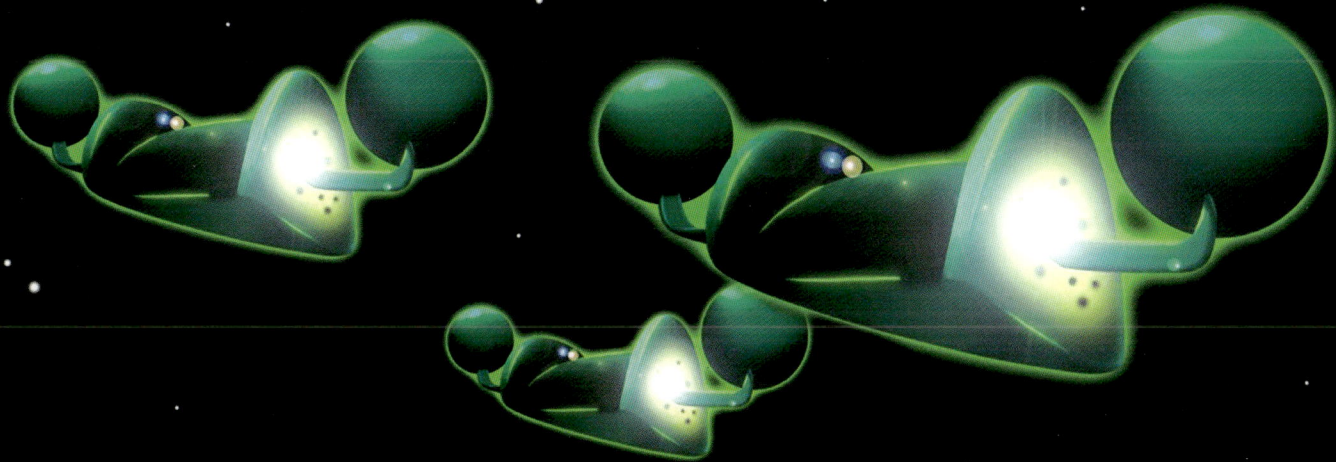

Emerald Trio

Jacky Kozan

While You Were Sleeping - On August 31, 1994, Andrew, Simon and Louise, from the rock group manmademan, were on Silbury Hill overlooking the West Kennet Long Barrow.

The three were in the company of five others who were enjoying the night and sharing in a few alcoholic beverages. By 1:00 am a thick fog had settled on the area, but not enough to obscure two orangey balls of light that were approaching from the left. Andrew, Simon and Louise turned to gage the reactions of the rest of the group, but noticed they all seemed to be "sleeping."

As soon as the balls were in clear sight, the orbs expanded and transformed into tetrahedral shapes which the three could see directly into. Inside each object were five or six small beings surrounding a taller being.

Louise grabbed her torch and flashed at the objects which, to their amazement, started to move towards them. The objects rose up towards the hill and moved to within fifty feet of their position.

As the three began to become fearful, the object moved back towards the field and hovered. Louise then noticed the beings from one object floating out of it. The beings stopped and hovered as well, and proceeded to draw a grid on the ground with a device that emitted light.

The grid was approximately one hundred feet in length on each side.Suddenly, the trio became aware of a vehicle approaching from the Marlborough direction.

The hovering beings also became aware of the vehicle and promptly retreated back into the tetra-hedral object, which then merged back with the other one. This singular tetrahedron then shrank back to an original orb shape and receded into the hedgerow, attempting to hide from the car.

Once the car had passed, the orb returned to its original hovering position and expanded back into the two tetrahedral objects. The beings re-emerged to produce yet another grid next to the original square drawing. It was approximately one hundred yards, only this one was rectangular.

Once completed the beings returned to the tetrahedral objects that then simply faded away. Oddly, when the objects were completely out of site, the five drinkers immediately "woke up!"

After initial investigation it was discovered that the car driver was none other than the late Paul Vigay, a well respected crop circle researcher, who had apparently witnessed nothing as he drove past the hill.

While You Were Sleeping   *Art by Kim Carlsberg*   *Andy Russell*

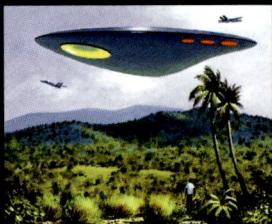

Puerto Rico Encounter - Near the Cabo Rojo area of Puerto Rico on December 28, 1988 - according to multiple eyewitness accounts - this gigantic Delta-shaped UFO was 'buzzed' by two Navy F-16 "Tomcats."

After the jets made several passes at this behemoth, the UFO appeared to slow as if stopping in mid-air. Then, witnesses claimed the two Navy fighters disappeared, as if swallowed up by this monstrous phantom. Shortly thereafter, with a flash of light, this UFO divided itself into two distinct triangle sections that flew away at high speed.

Both the FAA and the Navy denied any aircraft were lost or unaccounted for in the area, however, researcher Jorge Martin was able to confirm through a Navy source that the whole incident was monitored on radar, and that the radar tapes were immediately classified and shipped to Washington D.C.

Puerto Rico Encounter

Jim Nichols

Short Cut - I usually take a short-cut to work and back through the woods. It can be intimidating at night, but I'm OK with it... but yesterday I saw something so strange, I whipped out my camera phone and fired the flash.

Now the camera didn't pick anything up at all, but what I saw through the flickering flashes, went like this...

Short Cut   Tim Beeken

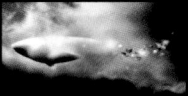

Through The Eyes Of A Child - My name is Michael Austin Melton. I'm pretty much normal and average. I was born in 1956, in St. Louis, Missouri, about a mile away from the great Mississippi River, on a street where the Budweiser Clydesdale horses were kept and groomed. I can't recall much from those days, except for the fact that life really began when I was eight.

It was likely the summer of '64. I recall looking out the back kitchen window that overlooks the slanted doors to the basement, and back yard. In the distance stood a giant apple tree, and the mimosa tree that grew up tall beside it. The sky was overcast, it was 7 o'clock, and I had just had breakfast.

Suddenly I noticed a sound, a deep hum, either inside my head or in the air around me. It then morphed into the sound of a hundreds propellers, or a powerful electromagnetic vibration. It was loud, but not painful, as I expected it to be as the sound grew louder. As I watched the overcast sky, transfixed on the cloud cover, I noticed ripples in the clouds moving across the sky like waves. They appeared gentle at first, but with the increasing din of the hum, they grew more and more storm-like. Then, what appeared to be a blimp or dirigible broke the surface of the clouds. It was smooth, long, and silver-gray. It dipped out of the cloud cover like an upside-down dolphin. As quickly as it moved out of the cloud cover, it slipped back into the layer of turbulent atmosphere. What followed is what I recall most: following the blimp, objects of every shape and size billowed out of the cloud bank, making an upside-down arc right back into the clouds.

My eight-year-old mind saw these objects as airplanes, all smooth on the surface, or with strange angles and curves. They moved gracefully, and I thought they were beautiful. Funny, at the time, I made nothing of the fact that I noticed no wings or propellers on the objects, but I had imagined them there. There were blinking lights: red, green, and blue, all quite brilliant and very metallic. As the last object passed through its upside-down arc back into the clouds, the loud hum softened. I could move my eyes again and noticed the trees. It began to rain. As the hum grew more faint, I turned to notice my aunt Ronnie in the kitchen. She asked what I was looking at, and I replied "Just some airplanes." She noted that it was going to be lunch time soon, and that I ought to run upstairs and get dressed, or I would have no time to play. I noticed the clock. It was 11:15 am. More than four hours had passed since I remembered staring out of the back kitchen window.

Soon after this event I took more of an interest in the planets and stars, the space program, and especially the night sky. I would always be looking up, perhaps for a reprise of that fantastic event that gave birth to my strong interest in things extraterrestrial. I mean to the point of running into light poles, and being the subject of laughter.

From that point on, I knew that we were not alone in this big, vast creation. God had indeed been generous and made life the rule in the universe. No matter how often science debates and doubts the existence of "spacemen" and UFOs, because of my experience that morning, I knew science had a lot to learn about the cosmos. What I didn't realize was that this little adventure, so many years ago, was a harbinger of things to come, and that I was destined to be a player in revealing the truth about life in the universe.

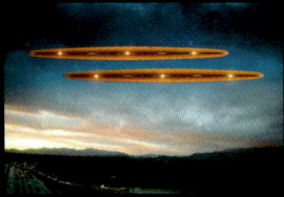

Bronze Beauty - On April 21, 2001 at approximately 8:00 pm, I was driving eastbound on Ventura Boulevard at Sepulveda in Los Angeles, California, and missed my turn.

I immediately turned around and stopped westbound on Ventura Boulevard at the stop light.

Approximately one half block away, I observed what appeared to be a hot air balloon gently "falling" from the sky. Then suddenly the movement stopped and what emerged were two large, thin, bronze disks, approximately three car lengths in diameter. I believe they were changing their altitude and direction of travel.

Their movements were calculated as they flew due north very slowly, in fluid movement, and in perfect symmetry with one another. They were following the course of the 405 Freeway.

I observed them for several minutes before they simply vanished. Although I only saw them briefly, the impact will stay with me forever. I now constantly keep my eyes to the skies.

Bronze Beauty   Art by Kim Carlsberg   Joliebeth Cope

Vasquez Visitor - On a crisp winter's evening in 1995, I was riding along with my best friend Chris, on the way to his house. We were traveling Eastbound on Escondido Canyon Road in Aqua Dulce, California, in the spectacularly beautiful region of Vasquez National Historic Park, along the Northern most border of Los Angeles County.

Vasquez is famous for a couple of things. All the old Westerns were filmed there because it has some of the most unique landscape and rock formations in the Southern California area, which are part of the infamous San Andreas Fault. The area is still the backdrop for many movies, television series', and advertisements. Still photographers adore the place.

Secondly, it is known as the official 'hold out' of Tiburcia Vasquez. In the late 1800s Vasquez, one of California's most notorious bandits, used the rocks to elude capture by law enforcement.

But Vasquez now has a new claim to fame in my own personal history. It is the place that forever changed my view of this little planet and our place in the universe.

On this particular evening, Chris and I were on the last paved road leading to the dirt road his house

was on. As we slowly cruised through the natural wonders of the rocks, we came upon a vision of elegance and beauty that left us both, speechless and breathless.

There, meandering in the night sky, with the comfort and familiarity of a dolphin in the deep blue sea, we observed a large, "electro-magnetheric blue", spherical object - resembling a comet down to the sparkling tail of light behind it. The sphere progressed fluidly not more than a mile ahead of our location.

I firmly believe it was an extraterrestrial spacecraft. It was reminiscent of a comet, but it was not a natural form. It was most definitely designed, constructed and maneuvered by an intelligence. It was huge, it emitted no sound, it moved gracefully but intentionally, it exhibited all of the characteristic typically reported of UFOs.

Fascinated by what we were witnessing, we continued to drive in the direction of the spacecraft for several minutes, before it simply dropped out of the sky like a falling star.

We raced in its direction at a very high rate of speed, trying to narrow down the location so we could get a closer look. We combed the entire area on all the narrow, winding, dirt roads but, unfortunately, we were unable to actually locate the craft.

As I have already confirmed, this incident proved to have a profound influence on my life. I have drawn pictures of spacecraft and pursued the UFO concept on a number of levels ever since.

Cylinder Ship - Over the decades, numerous eye-witnesses worldwide have claimed seeing gigantic cigar-shaped UFOs.

The origin and purpose of these craft remain a mystery. However, according to the captured records, during World War II, the Germans had construction plans for a 'Zeppelin' sized levitating cylinder ship called the "Andromeda Machine."

This 330 foot vessel was capable of carrying as many as three smaller scout ships.

In the early 1950s a California man named George Adamski photographed a UFO remarkably similar to this description.

Cylinder Ship   Jim Nichols

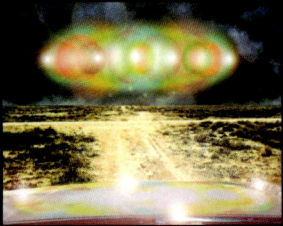

Shining Brightly - To this day I don't know the extent of the close encounter on that clear moonless night in the summer of 1970 on the high plains of central Wyoming. Late in life now, I wonder if it was an encounter of the third kind or even, possibly, a fourth because so much of it still remains unexplained in terms of reason, time, and possible life-changing after-effects. But that is another story.

Driving along a deserted dirt road near midnight on our way back home from town to a remote mining camp, I saw, far off to the south of us, a pulsing, lit object that changed colors, alternating from pastel reds, to ambers, to greens, to soft whites, back to red. The astonishing craft was flying slowly in our direction as we approached it. Our three children and our dog were asleep in the rear seat of the car and my wife dozed beside me. Fascinated, I drove on toward the craft as it flew, low and slow, toward us. I gently shook my wife awake, pointed out the object, and saw her instantly cringe with fear, shock, and awe. I kept her calm as best I could, trying to convince her that there was "nothing to be afraid of."

We met at a lonely crossroads where I stopped while the ship hovered, silent and motionless, a hundred yards in front of us. The craft, about forty feet in diameter, was oval but shaped more like a blimp than a saucer. It had a translucent, or diaphanous, appearance, no doubt because of those lights. So softly veiled, I felt I should have been able to see into it, yet I saw nothing, nor could I see windows or portholes. There was no sound that I could hear, but our dog began to howl, as if in pain, waking the kids and setting off their own howling at the sight of the ship filling the entire windshield of the car.

At a standoff, I decided to drive rapidly through the crossroads and on past the craft. Then, as if its navigator had read my very thought, it descended directly in front of the car as if to intentionally block us. It startling me and terrified my wife and the kids. Now, with no more than thirty feet between us, I felt the spine-tingling sense of the presence of living beings inside the ship. I still could not see through those diaphanous pastel lights, yet I "felt" that I could – as if it were no longer my eyes that were doing the "seeing", but rather another, unfamiliar, sense. Perhaps I felt my own first twinge of fear as well. We were trapped.

But then, after a seemingly short confrontation, the craft began to drift slowly Northeast above the crossroad, and our dog began to howl again. For several hundred yards it floated along, slowly gaining altitude and speed as we watched it, mesmerized. No one spoke; we simply watched it glide along for a mile or so toward the horizon of a nearby ridge, and then the ship flashed a burst of white light and accelerated at a rate beyond comprehension. In seconds it was a small ball of light the size of a full moon, then Venus, and then it was lost among the stars as if it were a star itself.

Shining Brightly Art by Kim Carlsberg Albert Lloyd Williams

Making Things Clear - In 1996, me and a friend were outside in my mother's car eating some cupcakes that she had brought home. I noticed a red light in the distance, but thought it was a cellular tower, (they had just started putting them up) but then it started coming towards us. So I told my friend, "Hey man, look at this shit!" ...and then it got about one hundred yards in front of us and stopped.

There was not a sound. It just sat there... a red light in the sky for about thirty seconds, and then it took a very slow left and once it got behind the trees we couldn't see it anymore. So I have been very interested in UFOs, aliens, and Roswell ever since."

Making Things Clear

Justin Lowe

The Orange Sphere - I had an encounter that changed the course of my life in 1947 when the family moved from Chicago to San Francisco.

I was 9 years old and going into the 6th grade. I cannot remember what month it was, but it was after July when we were in our first house.

It was a Sunday, we had just finished dinner and Dad, my younger brother Bob and I, adjourned to the living room and divided up the Sunday newspaper.

Mom remained in the kitchen cleaning up the dishes. We barely got settled with the paper when Mom called out to us urging us to come back to the kitchen.

There we saw an object that none of us had ever seen anything like before. It was a stationary, glowing, orange disc or a sphere. It was big!... bigger than the full moon. I estimated that it was not more than two miles away and it did not move, it stayed centered in our kitchen window. I was a self-trained plane spotter and prided myself in being able to

recognize any airplane made in any country. "What is that Dad?" I asked. He shrugged his shoulders. All four of us stood there in the kitchen and stared at it. I had the unshakable feeling that it was watching me. It seemed we only watched it for a few minutes until it just "winked" out.

I never forgot that incident and I can still see it very clearly, but the rest of the family quickly forgot the whole thing. From that time on, I read every UFO related book I could get my hands on (from that time until I retired I read over 10,000 books)

It was not until I joined a UFO study group in Santa Cruz, California that we got onto the subject of missing time and I realized that during that sighting there was about three hours of missing time.

Then I remembered a period of missing time previous to that in Chicago. Looking back, I can see that my life has been guided all these years and still is. I don't know by what or by whom, but there is no doubt.

The Orange Sphere   Art by Garth Perfidian   Lee Louden

Off Shore Encounter - When I was young, I had an interesting encounter on an empty New Jersey beach, during the off-season.

I was standing alone on the deserted shore near the top of a dune, about fifty feet from the road and about fifteen feet from the beach-front homes.

It was very cool and crisp that night and slightly windy. None of the houses in the area were occupied. I was in a world all of my own, or so I thought.

As I stood there appreciating the vastness and raw beauty of the ocean and sky, I was suddenly jarred from my contemplative state by a loud boom as simultaneously, "something" emerged from the clouds traveling at an amazing speed, then abruptly stopped in the sky in the distance.

Red, green and white lights traveling around the perimeter of the object pulsed in brilliant contrast to the blue/black night sky. It seemed as if "that something" was slowly traveling in my direction, when another, single - brighter - white light appeared on one side of the, what I then had determined, was definitely "a spacecraft."

The white light also circulated the perimeter of the craft as it continued in my direction.

At that point, I began to get a bit frightened and thought it would be in by best interest to get out of sight. I quickly turned and bolted toward the road behind the houses.

I felt somewhat safer hidden behind the vacant homes, but my curiosity got the better of me, I peeked around to look at the object.

Then, with a sigh of relief, and a twinge of sadness, I watched as the amazing machine traveled away from me over the sea... to who knows where.

Towering - When I was eleven, we moved from Burbank, CA. to New Hope, Minnesota for two years and then back again. In New Hope, we were out in the sticks, County Road 18, six new houses on our block and not many others for a quarter mile. About one mile away, just northwest from us was a tall water tower with the words Four Seasons on it, that was the name of that development.

Wide open spaces are just perfect for an adventurous eleven year old, life was good. On the back northeast corner of the property were two, fifty-five gallon drums that belonged to the neighbor, we could burn our trash back then which was a great pass time. One summer evening while burning trash, we looked toward the water tower to see something hovering maybe thirty or forty feet above it.

It was almost as wide as the tower which was, I'm guessing, sixty feet wide. The tower looked like a giant bagel on top of a single pole that was maybe twety-five feet wide. The hovering object had lights going around it. I remember my dad saying "Wayne go get the binoculars" and when I came back out again my parents and my sisters were all standing at the property line staring at the object. That is the last thing I remember about being in Minnesota. I have just a few memories about the trip back to California. Understand, I have a very clear memory of everything that happened up to that point. Neither of my sisters remember anything about the UFO or the move back to California.

Years later, I noticed a spot on my right eyebrow that is very sensitive when I touch it, but if I don't touch it I don't notice it. Additionally, over the years I have had reoccurring dreams. In one dream I look up at the sky and it is covered with pretty white clouds and I can see blue sky but very clearly on the other side of the clouds the sky is filled with UFOs of all different types. My dream continues with me finding my way through town hiding and ducking behind things... it's always the same.

I was only eleven years old at the time of my sighting and I don't remember everything, but of what I do remember I have a good recollection. Recently, I am remembering more. This morning I woke up too early to get up, so I figured it was a great time to meditate. During the meditation, I had a vision of being back in Minnesota, coming out of the house with the binoculars, and seeing my family standing in fixed positions like statues. I then continued walking toward the water tower and when I made eye contact with the craft my temples pulsed several times and then my vision and consciousness went blank.

For several reasons, I sincerely think I was "taken" at that moment, and I also think my families memories were erased. I cannot tell you the time frame from that event and when we packed up and moved back to CA. I have never claimed to be an abductee, I really don't have much of a story but it is very real to me, and I seem to be waking-up to more and more details.

I can't look at the sky without expecting to see something, and frankly I am disappointed most of the time. I believe my "frequency" is rising... I am constantly seeing things out of the corner of my eye, and it's happening more often. I feel like the guy on Close Encounters that is always seeing the vision of Devil's Mountain. I know something's up, but I just can't pin point it.

Towering   Art by Andrew Pearce   Wayne Miller

Silver Sighting - In July of 1997 I was driving alone on Louise Street just north of Nordhoff Street in Northridge, California in the mid-afternoon when I observed a silver, triangular-shaped, spacecraft approaching my vehicle.

The craft neared my vehicle then came to a stop and hovered over my car approximately six feet above me!

I noticed three circular lights underneath the craft directly in the middle in a straight row. Two were green and the one in the middle was white, each one was about the size of a baseball. The craft was just smaller than a mid-sized automobile. There was no sound.

The craft moved very slowly while it was approaching my car and hovered long enough for me to get out of the car and for the occupants of several other cars to slow down and stare along with me in complete astonishment at the craft.

I was so excited because I had seen other craft at a distance before, but this was absolutely without question a UFO sighting, UP CLOSE, IN BROAD DAYLIGHT! It appeared to be a deliberate attempt on their part to communicate their presence to me, since there were other cars around me and it was directly above my car.

Although the approaching movement was slow and deliberate, its sudden departure was so rapid it was almost inconceivable.

I was elated at the event, because I had observed what I thought was a craft following me at a distance on many occasions over the previous seven months.

Silver Sighting   Art by Kim Carlsberg   *Joliebeth Cope*

Celestial Celebration - I am driving through the Mojave Desert alone in my car, following my boyfriend who is alone in his car that he picked it up earlier in LA. We are both headed back to Tucson. It's early evening.

Shortly into the trip, I see this extremely vivid star in the sky that keeps getting brighter. I'm thinking, "This is interesting!" Instinctively, I slip my "Close Encounters of The Third Kind" 8-track tape (we're in 1977) into the deck. I keep repeating the "Dah da Dah da Dah" song. I'm wishfully thinking, that if I play this music and the strange orb is an ET ship, the occupants will be able to detect the music vibrations emanating from my car. Surely, any visiting ETs will know **this** song!

Suddenly, the radiant light goes out as if a switch has been flipped... poof!... it simply vanishes. It didn't go behind a cloud, because there are no clouds! Then, about fifteen minutes later, I notice it reappearing again. It very slowly increases in brilliance. This is really something!... so I repeat the music sequence again and send them a call from my mind. I say, "if you are there let me know it, I really want to know, if you are there let me know it." This

dazzling star then, once again, pops out of sight but almost instantaneously an object slowly and gently descends upon my car like a feather dropping out of the sky. It falls softly and with no sound. It is oval shaped, but instead of having rounded edges, they are squared off. I'm perceiving what I think is the bottom, but I can't be sure.

There are hundreds of rows of white lights reminding me of small Christmas tree lights. I'm thinking this is so fantastic, but not the typical UFO one would expect to see. I am really excited that they are responding and I desperately want to pull off the highway. I start blinking my lights and sounding my horn, hoping my boyfriend ahead of me will stop, but he just keeps going. I didn't know my way back to Tucson, on my own, so I decide to keep following him rather than pull over.

Whatever it is, it is less than 100 feet above my car, making itself clearly within my eyesight. I feel very disappointed because "they" must have heard my call... they are here, and I am not stopping. I say to them "I must keep going... I'm very sorry... I have to follow my boyfriend back", as I sadly drive past the object. When we arrive in Tucson I ask my Boyfriend, "Did you see it?" I am still very energized. He looks very dazed and perplexed and says to me, "I don't remember driving back, it's like I was asleep at the wheel. I don't remember anything!

My art is not so much channeling, but tapping into the unified field for like energies. These humans are in our timeline coming from the past, but they partake in our distant future. Their names are Anzara & Sondavidias.

Celestial Celebration

Charla Gene

Blackout Days - Santa Clara, Cuba (a city in Central Cuba) 1965 around 2000-2100 hours. It was around this same time during one of the numerous blackouts in the area that while several neighbors sat outside talking, I played Dominos with some of the other children. We looked up as luminous object seemed to drop out from the sky and glide above a row of houses directly across from mine.

Of course I had no idea what it was, and I had never heard of UFOs or Ovnis. It seemed to be a metallic egg-shaped craft which emitted a luminous beam of white light from its bottom section. No one could hear any sound coming from it and many of the neighbors yelled out that it was "Los Americanos!" due to the post - Bay of Pigs - hysteria that persisted in the region.

After a few seconds, the object glided over my house and over the indoor patio. I ran inside the house and

ran after the object but as I looked back I realized that I was alone... everyone else seemed to have lost interest on it. The object stopped over the patio and seemed to hover there forever while I stood mesmerized under it, staring at its lights. Before I knew it the craft was gone and it was very late, I seemed to be wet and everyone had gone to bed.

I have no other memories of that night. Also, around the same time period, again while under another blackout, I was outside playing alone in the street when I heard a terrific sound directly above me. Today I could only compare it to an aircraft breaking the sound barrier, that very well might have been the case.

A few days later, again under blackout conditions, several neighbors and I saw a huge 'ball of fire" hovering close to the ground at the end of the unpaved street... we all ran into our homes!

A Saturday Afternoon - It's a hot September Saturday, with blue skies full of sun and plenty of green as we are in the countryside. It is a lovely day. Red Hot Chilli Peppers are playing on the stereo of a maroon B.M.W. 3 Series that houses all the trimmings from the walnut dash to the cream leather seats. It is a convertible with big, wide, sporty alloy wheels... the hood is all the way down. The car is in motion along the roads of Britain.

The driver is an Italian looking guy, tanned and toned with a chiseled jaw, a little stubble on his chin and short spiky hair. He goes by the name Beeks, his nick name since secondary school. He is wearing dark expensive Bolle sunglasses, a sleeveless black t-shirt and jeans. Beeks glances in his blue tinted mirrors, there are some cars sparkling in the distance on the same route perhaps. With the road clear ahead Beeks presses the accelerator and a deep powerful roar bellows and he glides along the road to pick up Rob... they are going to Camber Sands, a stretch of coastline south of Kent.

On entering a small village, Beeks pulls off down a long, slightly dusty track with tall trees, to nothing short of a white mansion with pillars, a massive gravel drive, and statues. The door to the big house opens and Rob appears. Because these guys are **cool,** a quick nod from both suggests hello! Pale

Rob stands taller than Beeks, has a normal build and pleasant face. He is wearing black shorts, white socks, yellow trainers, a black cap and a snug, black t-shirt that reveals his tattooed arms. He's got some CDs and his wallet in his left hand as he walks up to the car and smirks "How's it going man?" They do a "cool" secret hand shake as Rob slides into the passenger seat and they both grin thousand dollar grins as they set off to, once again, luxuriate in their good fortune.

As they pull into the village centre, there are many people sitting and walking around in the park, all in summer attire. As the guys drive through, they notice a small amount of congestion has built up due to some parked cars on the sides of the road. The road is downwardly steep as they neatly exit the village and there is great scope across the valley, an effect of the contoured ground... you can see far across to the other side that is full of hills and Kentish scenery.

Along the way the two exchange gossip and thoughts on certain subjects, just general banter... nothing too deep. On entering the main road to Camber, Beeks presses the gas and the deep bellow and whistle of the 300 bhp engine nudges the car forward at an unrelenting speed as they enjoy a short burst of adrenaline filled freedom. Rob suddenly notices gasped faces in the other cars that they cruise past. Camber is not half a mile away so Beeks eases off the accelerator. Camber is written on the next exit lane sign. Tick, tick, goes the left indicator as the car pulls off round the sharp bend, the grip of the wide wheels causing no alarm. "BEEKS!!!" yells Rob... "LOOK... something is wrong with the clouds!!!"

A Saturday Afternoon   Tim Beeken

Burning - August, 1996 - Makerfield Way, Ince, Wigan, England.

At approximately 10:05 pm, my wife, Erika, and I left the clothing warehouse after a full work day, and started towards the bus stop down the road. We were hoping to catch transport back to our nearby, hometown of Atherton.

It was a clear, windless, enjoyable evening. As we walked, our eyes fell upon, what we thought was, an extremely bright star. Suddenly it started to glitter brilliantly and move towards us. We watched in amazement as the sparkling light approached, and then stalled in the sky over the nearby fields. It hung motionless momentarily before abruptly bursting into a burning, fiery, orange/yellow mass.

Fascinated, I ran towards it... though my wife was scared and cried out to me to stop. I have been interested in the UFO phenomenon all my life and I wasn't about to miss any of this, despite my wife's protestations!

We then observed two helicopters approaching from opposite directions, both racing towards the craft. We could see no lights or markings on either

of them. As they got close to the object, we watched the incredible, broiling celestial body suddenly shrink into nothing and seemingly vanish. We then noticed it slowly re-emerge in another part of the night sky, back to its full, blazing form.

The helicopters immediately turned and made towards it again. As they re-approached the sphere, it once more dissolved and, seconds later, reclaimed its original position. A third time the helicopters moved after it, and instantly it blinked out and popped back into view further away. It was a game of cat and mouse with the UFO quite clearly in full control of the game!

Despite my proximity to this activity, my feelings were not of fear or apprehension, but exhilaration and amusement. Amusement in the way the craft so easily eluded the helicopters I should add. At this point the helicopters gave up and flew off. We last saw the flaming globe descending behind the trees several fields away.

Had the distant not been so great and insurmountable, I would have willingly gone after the UFO, even if it meant walking home that night. My wife felt only fear though, and urged me to go with her since it was already out of sight. Reluctantly, I agreed.

I've had other sightings in my life, but this was, by far, the most spectacular. It left me with no doubt that the astounding object was intelligently controlled and from beyond either this Earth, or this dimension... or both!

Shortly after, I contacted a local UFO investigator who told me the Civil Aviation Authority claimed that nothing was in the sky at the time. So who were the helicopters?

Burning

Art by Kim Carlsberg

Mike James Gorman

Light Beams - Haravilliers, France (30 miles from PARIS) January 10, 1998.

By 7:30 am, seven witnesses were driving to a hunting expedition in two cars. The weather conditions were exceptionally clear and many stars were visible.

As the two cars were approaching the parking area at the hunting preserve, their occupants saw through their windshield a bright illumination in the sky. A huge black round object (ten to fifteen feet in diameter) comes towards them, slowly gliding only about seven feet above the ground.

Mr. D notices there are a series of six to seven light ramps bearing red, green and yellow spotlights. The beams do not light up the surrounding area. They resemble those of a Christmas tree decoration. Mr. D continues to approach the location where the object was hovering. As the car went just beneath the object, he could see a dull gray platform with a bright octagonal in the center. A beam of light came down to the ground, and then the testimonies saw three green, red and yellow beams converging on the Mercedes.

Later, around April 20th, Mr. D's memory begins to return little by little. Mr. D recalls having been levitated into the hovering UFO through the central hole. Inside he was confronted by a humanoid figure about four feet high, completely covered in something similar to a diving suit, his face was hidden behind a metallic, cylindrical, dull-black helmet.

Extracts from the report of Gerard Deforge published in the Magazine "Lumières Dans La Nuit" Nr 352 - June 1999.

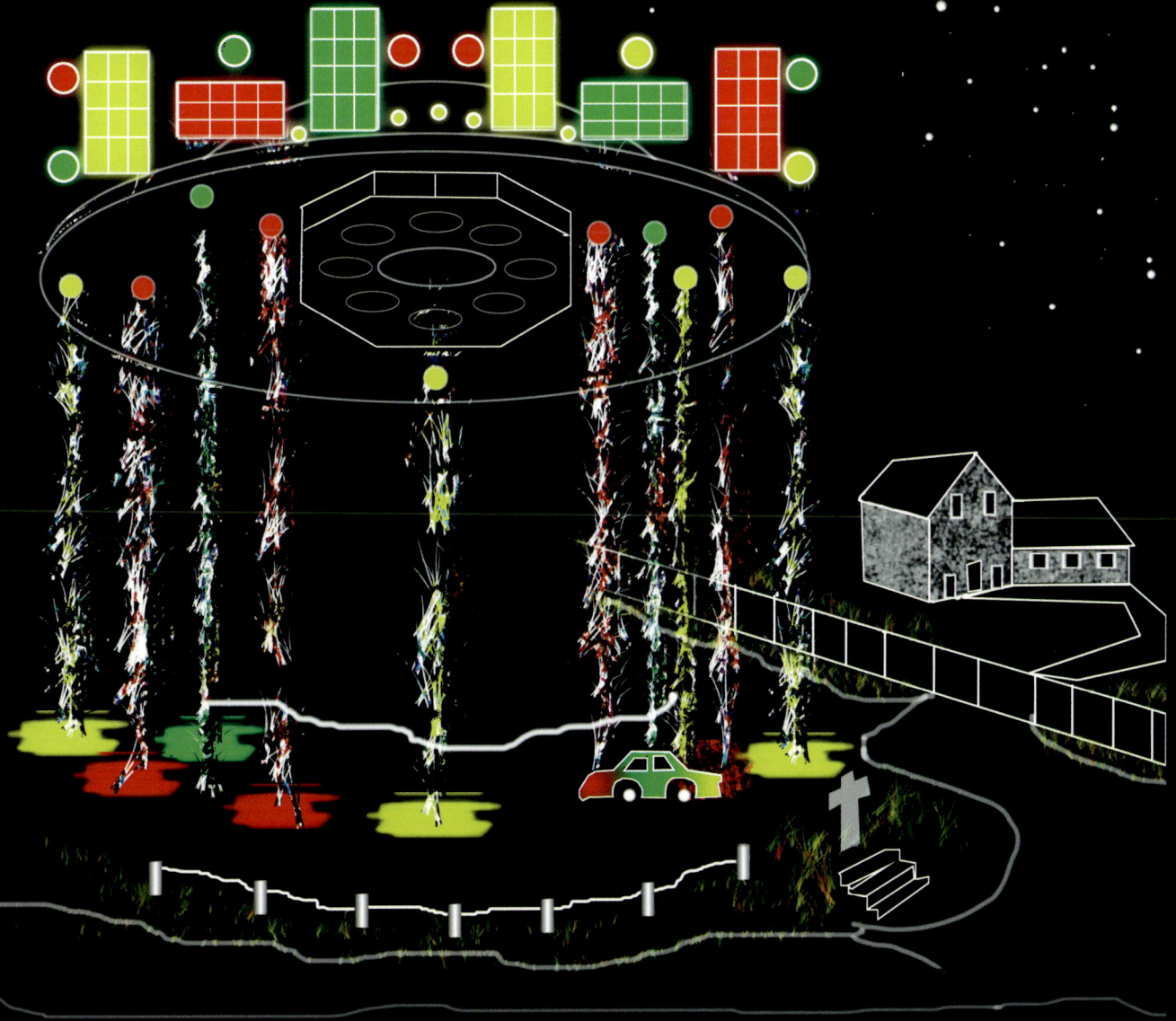

Light Beams

Jacky Kozan

Area 51 Triangle - It was March 27, 1996, I was living in Las Vegas, NV. Having an interest in UFOs since my childhood, I took advantage of living that close to the Area 51 base and would visit often in hopes of seeing something.

Normally I'd plan my trip days in advance, however, that night my gut was telling me I had to go. This gut instinct or intuition is something I've experienced regularly since my teens and had never proven me wrong, so I took it seriously.

Coincidentally, I was sent home early that night as business was slow. I informed my wife of my plan, change clothes and headed out the door. Soon after, I was on Hwy 93 coming around the first big bend in the road. Las Vegas and the interstate quickly disappeared behind me. It was around 10:00 pm and I was the only car on the road, until I got completely around the bend.

I soon noticed another car heading my way off in the distance. It seemed normal at first, but as I got closer I realized the car wasn't moving. This worried me as would try to stop me. My worries quickly turned into fear when I realized the lights were actually hovering over the highway. I immediately slammed on the brakes thinking if might be a Police copter... I was going way over the speed limit. After thinking about it for a few seconds, a Police copter

out there that late seemed ridicules to me. That's when I realized the lights up ahead were why I was drawn there. I stepped back on the gas in hopes I wouldn't miss it before it left. I grabbed my binoculars and tried viewing the object while driving at high speeds. This was nearly disastrous so I put them back down. I did see enough to tell it was an illuminated solid object over the highway. It stayed there until I arrived and parked my car under the front end of it.

I could see it was a decent sized triangle, maybe four car lengths front to rear. All sides were equal in length. It wasn't black though, but more of a dark, charcoal gray. It was about twenty feet above the road. The underside had what appeared to be nine large, yellowish/white, filament floodlights or incandescent bulbs. It was so quiet all I heard was the wind blowing through the brush and the slight hum of my car engine.

The craft hovered for several minutes as I studied the bottom of it, though the lights were really about the only details I could make out. I reached for my binoculars for a closer look, but when I turned around, the triangle was gone. It had shot off into the distance, now at a lower elevation as if to allow me a better view of the entire craft. The triangle was actually a short tetra-hedron. Its flight followed the contours of the ground until it eventually disappeared over the mountain range to the north.

I looked at the clock and it was 10:10 pm... no missing time. I continued onto Area 51 after the encounter, but nothing else eventful happened. The return drive home was without incident as well, other than a flood of questions and emotions.

Area 51 Triangle

Shawn Jason

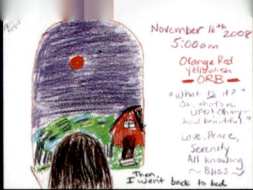

The Orange Orb - On the weekend of November 15th to the 16th, 2008, my husband drove out of town with our young son to visit his grandmother who was turning 101. I had to stay behind to care for my hospice state dog who was 16 years old: blind, deaf, incontinent and needing special supervision.

I was excited because this would be my first night alone, by myself, since becoming a mom four years earlier. That Saturday night, the 15th, I went to work out at my gym, came home, had a quiet meal, then drove to the Borders and was back home by 11:30 p.m. I was a little anxious, but mostly about burglars. I turned off the light at 1:00 a.m. after getting my dogs settled into the bathroom next to the bedroom I was sleeping in.

At 5:00 a.m. (before sunrise) something stirred me. I thought it must have been my older dog bumping around, needing to go out to potty. I jumped out of my bed, heart racing for fear of him letting loose on the tile, only to find... he was fast asleep! "Huh... that's strange" I thought. I could have sworn I had heard something. Immediately, I felt compelled to look outside the front door window panel. I told my self I was just curious to see how light the sky was since I'm

rarely up at that hour. When I peered out the window of my front door, I was absolutely astonished to see... there in the predawn, purple-blue sky... a warm-toned, reddish-orange colored Orb suspended in the sky across the street above my neighbor's house.

It was not the sun. The sun hadn't risen and wouldn't for another hour and a half at least. It wasn't a planet, it was too big. I stood there, shaking my head and thought, "What... is... that?" A moment later another thought entered my mind, "Oh, that's a UFO!... Oh my, how beautiful!" I realized instantly I needed to take a picture, but remembered that, unfortunately, my husband had taken the camera out of town with him! Sadly, I had no way to document the event, and coincidentally, no one to share it with!

As I stood there admiring the mystery presence, I became entranced by it. I warned myself to be cautious, but I was absolutely enamored! I felt a powerful, all encompassing, serenity from the craft. It projecting warmth, tranquillity, peace and love... no threat or malice whatsoever, only benevolence. Physically, and energetically, it was absolutely gorgeous!

But even with this enormous flood of blissful feelings, I understood I was witnessing "an unknown" from another world or realm, and I was glad I was seeing it from the safety of my home. I was aware enough to deny any possible urge to venture outside my door. Then, after an undetermined amount of time, I simply shrugged my shoulders, and... (this is what puzzles me most)...went back to bed...?...!

November 16th 2008
5:00am

Orange Red
Yellowish
— ORB —

"what IS it?"
Oh, that's a
UFO! Oh my —
how beautiful"

Love, Peace,
Serenity
All knowing
~ Bliss ~

Then,
I went back to bed

The Orange Orb
Melissa Kriger

Overpass - My first career (as I like to call it) was a professional, cross-country, hitch-hiker. I covered the country back and forth many times and had adventures in every state, but this is one of those stories that will never fade from my memory.

I was traveling through central northern Florida at 3:00 am where the I-75 north crosses the eastbound I-10. I was headed west towards the Alabama border. My ride was continuing north, so our time together came to an end. I was dropped at a lonely off-ramp quite a ways outside of Lake City, but I didn't look for a ride right away, I kicked back against my trusty duffel bag to "enjoy the journey" as they say, and have a smoke. It was early fall and the days were mild and the nights were crisp and unclouded.

I was contemplating calling it a night and making camp alongside the road under the moon shadow of a Laurel Oak. As my eyes scanned the landscape searching for "my energy spot" as Carlos Castinada coined the term... I noticed a pulsating band of neon, pastel colors hovering very low over the tips of a stand of pine trees off in the distance to the north.

The night was clear, and as I watched in astonishment at the noiseless object moving slowly in my direction, I was able to make out the shape of a giant metallic disk cocooned in the illumination. My heart started beating quicker. I was no longer languishing in my mellow journey, I was instantly hyper aware and very excited, my eyes transfixed on the silent, gliding carousel of color. There was suddenly a suspicious absence of passing cars. I looked behind me, only to see nothing, and no one. I wasn't exactly frightened, just very unsure, and I was totally bummed I had no one in which to share this "once in a life time" event.

The craft continued its approach in silence but grew larger almost exponentially with every yard. When it was less than forty yards, and this is no exaggeration, I estimated it to be the size of a typical K-Mart parking lot. I noticed my mouth was parched and I realized I was now sitting very erect but hadn't summoned the courage to stand. I suppose I was becoming as afraid as I was stoked to be witnessing what could be none other than a giant UFO! The closer it got the more detail revealed itself. There were rows of small illuminated windows silhouetting "figures" inside.

I thought for sure it was coming right for me when, at the last twenty yard line, it casually turned and crossed the freeway at a snails pace approximately forty feet off the ground. There were four lanes in each direction... plus the emergency lanes... plus a center median and when the ship was finally centered over the freeway it still overlapped both sides. It eventually made its way across and floated down behind the tree line. I knew it had landed, and I also knew that I had had enough excitement for one night and hitched a ride in the opposite direction.

Note: I had described the ship to the artist but not the immensity. This ship is very close in description, but triple the size in your mind's eye!

Overpass   Art by Kim Carlsberg   Kevin Reid

Bashar's Craft - On two occasions within the same week in 1973, I had close-range, broad-daylight sightings of UFOs with witnesses present both times. At each sighting we saw a dark metallic, triangular craft about 30 feet on each side. There were three, blue-white lights, one on each "point", and one orange-red light in the center. The craft in the first sighting was about 150 feet away; in the second sighting, only about 60 feet away.

After seeing something that I knew could not be the product of Earthly technology, I was curious to find out all I could on UFO phenomena. I began reading everything I could find. Browsing bookshelves for UFO literature, I quickly discovered other books on different "paranormal" subjects, such as psychic powers, spirits and channeling. I read a few of these as well as UFO books so that I could broaden my research and acquire a greater understanding of the metaphysical field of knowledge.

Ten years after the UFO sightings, I was introduced to a practicing channel. After several months of listening to the information delivered by the spirit entity being channeled through the person, I was amazed by the consistency and quality of the information I was hearing on a variety of subjects. Eventually, that entity offered to teach channeling to whoever wished to learn. This surprised me at first, as I had assumed channeling was not something that could be taught. Nevertheless, I joined the channeling class — not intending to become a channel myself, but rather to learn more about the process by which this entity seemed capable of accessing volumes of information on endless subjects. Midway through the course, which contained many guided meditations and mental exercises designed to shift one's consciousness to various states, I received what sounded like a telepathic message in my mind. I became instantly aware of three things: The message was from an extra-terrestrial consciousness that I was to call "Bashar" (the UFO I had seen was his ship), a memory came back to me that I had made an agreement at some point prior to this life to channel him, and that now was the time to fulfill this agreement if I still felt like doing so.

At first, I questioned this internal experience: Was I hallucinating? Was this some strange "side effect" of the meditations the class members had been given? However, while I sat silently pondering some of these questions, the entity teaching the class became aware that I was communicating with something from another plane of existence. He urged me to trust and learn to work with it.

After thinking about it for awhile, I decided to explore the possibility of letting this "Bashar" entity speak through me to see what would happen. I figured that even if it wasn't really another entity — even if it was some mysterious portion of my own consciousness — the information that could be accessed through this channeling process could be used to help people make constructive and positive changes in their lives. Whatever the source, I decided to continue. I have now been channeling publicly since that time in 1983 and "Bashar" has spoken on a wide variety of subjects to thousands of people in more than fifteen cities throughout the United States, as well as Japan, Australia, New Zealand, Canada, England, Egypt and Greece.

Bashar's Craft   Art by Kim Carlsberg   Darryl Anka

125

The Guiding Light - Stonewall, Manitoba Canada, 11-'76, at approx. 0530 hrs, on an early, brisk winter's morning, my three siblings and I were unexpectedly awakened by my father... my mom was already up. My dad was a policeman and always aware of his surroundings, but little did I know that, on this particular morning, his well-developed skill would change the course of my life and the way I would think... forever!

My father's voice trembled with excitement as he shook us, one at a time, demanding we hurry to look out the back picture window of the living room. With confusion and hesitation, we all crawled out of our beds, stumbled into the living room, and laid our sleepy eyes on an object of such beauty and brilliance, it was hypnotizing. This light was as bold as a nearby planet or star on a cold cloudless night, with dazzling points all around it. It had no particular shape, only brightness like the image of "The Guiding Light" in the well known Bible story.

The object sat still for a bit, as if waiting for its audience to assemble, before it gave way to an awesome aerial display. My father eventually informed us, earlier that morning he had been warming-up his car when the object appeared out of nowhere and proudly performed for him. I'm guessing he watched it for a few riveting minutes before he realized he could not experience this alone... anyway "it" did wait until we were all huddled together. As we stood there... literally in awe at the brightness alone, the craft suddenly moved to the east at a high rate of speed, then abruptly stopped.

A moment later again, with instant acceleration, it moved to the west and again promptly halted. We all wondered aloud what is was, and then, as if on cue, it did a 360 degree circle a couple of times before suddenly stopping again. This glowing acrobat was performing feats that defied logic and gravity. This went on for some time... basically at the same rate and remaining the same luminosity, I'm guessing for ten to fifteen minutes until it instantly shot away to the north, again at enormous speed and disappeared into the heavens. We were all so excited, and surprisingly, not frightened at all, by the mystery we had just witnessed.

There was nothing else we could do but go back to our normal lives. My dad went to work. I went to school and told some of my friends of the morning's fantastic event... they were enthralled. I continued to scan the sky the rest of the day and on my way home from school... but noticed nothing out of the ordinary.

Now, in hind site, I realize that event radically altered my perception of the universe, and my life. Although I grew up in an agnostic family, I am now a very spiritual person, but my belief system is uniquely my own. And though I was not raised Christian, I do believe in the Bible and revel in its stories. I am also sure much of my spirituality stems from that special sighting in my youth. I am 41 now and I've been handed my share of challenges. I believe someone, or something, knew the power that sighting would have over me... I believe it was a gift.

It provided me with proof of at least one form of outer-worldly higher intelligence, which suggested a larger universe on the whole. I now have a strong, long standing relationship with a higher power in my life, and I believe that it was shaped, in large part, by that one, cold, miraculous morning... in November of 1976.

The Guiding Light   Art by Kim Carlsberg   Randy Pull

Abduction Encounter Art

Fire In The Sky - On November 5, 1975, the work crew I was on witnessed a spacecraft from another world hovering silently between tall pines in the Apache-Sitgreaves National forest of north-eastern Arizona. Subsequently I became an unwilling captive of an alien race when the other men fled in fear.

The story of my disappearance instantly became the focus of controversy and national attention.

When we stopped to study the craft, I left the safety of the pick-up truck and ventured close to the saucer-shaped craft, which eventually emitted a beam of light that struck me and left me unconscious on the ground. Panicked, the remainder of our forest service contract team fled. When they reported an encounter with a UFO, something they would have considered impossible if they had not witnessed it themselves, the men were suspected of murder. For five days authorities mounted a massive man hunt in search of me, or my body.

Then I reappeared disoriented and initially unable to tell the whole story of the terrifying encounter. When I recovered, I had a story to tell that was unbelievable.

I awoke inside a room on a table. I looked at the vague but reassuring forms of the doctors around me. Abruptly my vision cleared. The sudden horror of what I saw rocked me as I realized I was definitely not in a hospital. I was looking square into the face of a horrible creature... with huge, luminous brown eyes the size of quarters! I looked frantically around me. Hysteria overcame me instantly. I reacted in reflexive fear, lashing out at the non-human entities clustered around me. They stood still, mutely. They were a little under five feet in height. They had a basic humanoid form. But beyond the outline, any similarity to humans was terrifyingly absent.

The only facial feature that didn't appear underdeveloped were those incredible eyes! Those glistening orbs had brown irises twice the size of those of a normal human eye's, nearly an inch in diameter! The iris was so large that even parts of the pupils were hidden by the lids, giving the eyes a certain catlike apparentness. I would eventually encounter completely different, human-like beings who enigmatically treated me as though I were a frightened animal.

Two men and a woman were standing around the table. They were all wearing velvety blue uniforms like the first man's, except that they had no helmets. The two men had the same muscularity and the same masculine good looks as the first man. The woman also had a face and figure that was the epitome of her gender. They were smooth-skinned with no blemishes. No moles, freckles, wrinkles, or scars marked their skin.

I wrote a book about my experience 'The Walton Experience', which, along with the illustrations by Michael H. Rogers, became the best documented account of alien abduction ever recorded.

In 1993 Paramount Pictures released a movie, 'Fire in the Sky' from a Tracy Torme script that artfully examined the personal drama that ensues from an abduction event. Unfortunately, studio intervention prevented an accurate recreation of the on-board events, but the human side was presented intact.

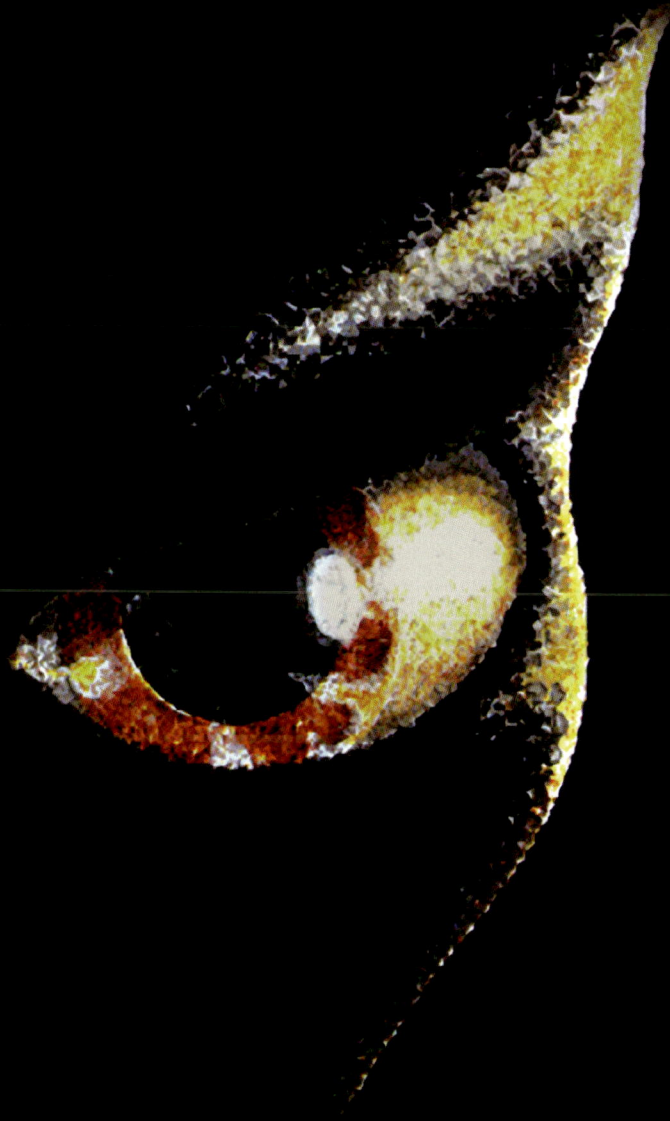

Fire On The Sky   Art by Michael H Rogers   Travis Walton

Bedroom Window - This illustration is an attempt to convey a very short memory. The incident lasted, maybe, less than twenty seconds. The implications of this event have been difficult for me to integrate into my life. I've been terribly conflicted about the truth of this foggy incident. It could have been a dream, true enough. I cannot allow myself to discount that.

In January or February of 1993, I was living alone in a small house in rural Maine. I had been dealing with a hard break-up, and I was in a fragile, emotional state, and this is an important factor in this story. The driveway of the house had one of those motion sensitive lights, and it was pretty common for the light to come on when a car drove by or a deer walked through the front yard at night. The bedroom window faced the driveway, and my bed was up against that window.

I was depressed during that chapter of my life, and sleeping poorly. But I woke up that night because a very bright light was shining in through the window above the bed. I sat up, propping myself up on my elbow, and looked out the window. I saw five spindly entities with skinny bodies and big black eyes. They were on the lawn walking toward the house. They were back-lit by a singular round bright shape.

This light seemed oddly small. My response to this frightening image was to nonchalantly lay my head down on the pillow and promptly fall back asleep. Shouldn't I have jumped out of bed screaming in terror? But instead I felt absolutely empty of emotion. It was almost as if I was somehow controlled. I calmly thought to myself, "Oh yes, they're here, let's just shut down and black out." The illustration seems to capture the memory as close as I possibly can. Now, its important that I add this extra information. This memory is strangely vivid in a way that seems entirely different form a normal state of mind. I saw something, but at the same time I truly do NOT think it happened in "this" reality. That may sound hard to grasp, but it is the only way I can honestly depict the experience.

It's important to me that I try to describe this feeling. Weirdly quiet, sort of a pressurized fish bowl, the deepest part of my psyche is displaced, and moved to the forefront - maybe the normal thought chatter in my head is turned off - maybe - kinda - sorta... a distinct warping of my psyche (whatever that means). Did it happen while sleeping, and I simply imagined everything? Was it just some sort of dream state? Maybe. That would be an easy way to sum it all up. I'll add that, because of this strangely vivid state of mind was so weird, I do not fully trust this memory.

The next morning I had the image seared into my mind, and I could NOT imagine something so weird could be true. I simply dismissed it as some sort of dream. I will add that I don't recall ever dreaming I was in my own bed. My dream imagery will always be somewhere else, never in surroundings that are exactly like my bedroom.

Bedroom Window

Mike Clelland

Tall Whites - One night, while sleeping with a companion, I abruptly awoke instantly thinking something was wrong.

As I sat up straight, I became immediately entranced and transfixed at the most amazing sight of four humanoid creatures standing at the foot of my bed.

There was a very small, bluish-gray being about three foot tall on the left side of the bed, while another shorter one of tan color stood on the right.

Two very tall, white beings, similar, but not as disproportioned as the smaller ones slowly approached behind the shorter creatures. They were all hairless. The taller entities had slanted, black eyes that were much smaller than the eyes of the tiny beings, but still a bit larger than human eyes.

The "talls" skin was paper white, their limbs extremely elongated, and their hands had six spindly fingers. While digesting the incredible scene, my field of vision narrowed and I physically felt my consciousness slowing, as if I was shutting down.

At that moment I mustered all the energy I could to repeat and remind myself... "I am not going to give up... I will remember this... I will remember you." As I fixated these words in my mind, I forced myself to continue looking at the tall whites. One approached with a cylinder object that literally floated in front of its finger. As the tiny tan being neared me... I felt a nudge on the right side of my head, and all went black.

That morning, I woke to find five needle-like dots in the webbing between the large and second toe of my right foot.

I immediately explained to my companion what transpired during the night. She was mesmerized, as she had heard of these beings, but had never seen evidence of their actual existence. We took photographs later that day.

Following this encounter, I didn't know if I had been implanted with anything or not. I actually felt a bit special somehow... privileged at having this type of experience... disturbing as is was.

I did consider having x-rays of my right foot, but doubt and fear stopped me from completely accepting the encounter as a reality.

To tell you the truth, the event was so scary I wasn't sure I really wanted to validate it, or the reality of the visitors. Instead, I choose to deal with it through my art.

Tall Whites   Christian Fedor

The Innocents - Paranormal activity began in my life about two years ago at the onset of a series of UFO sightings. These events have dramatically changed my life. I didn't know what to do in the beginning, other than report them, but I soon realized there was just no point.

I recall at least five other UFO incidents, but I recently concluded that these sporadic sightings were just the tip of the iceberg of what was actually going on with me.

I started connecting the dots and realized that much of the "high-strangeness" I was experiencing, could very likely be directly linked to the UFOs and their occupants. For example, I am 32 years old and suffer night terrors. When I recently researched the subject, I found this is something that usually only affects young people... until I further discovered it is also a phenomenon commonly reported by people who claim to be "abductees."

A typical "paranormal" experience for me is waking up completely paralyzed to the extent I can barely manage a moan. I am always aware of a presence. I have felt my legs thrown apart even though I couldn't move a muscle... and there are usually "owls" outside the windows, observing... waiting.

A hazy memory of an Asian featured female being inserting something like a tube inside me haunts me. Nosebleeds have became a common occurrence and I started discovering lumps and lesions inside my nose.

I have awakened with a triangular scar, scoop marks and with my clothes inside out and backwards.

One of the more frightening memories was like waking up during a surgical procedure. I was aware of physical manipulations being performed on my body, but I couldn't move or talk. At one point, I actually thought I was dead and being worked on by a mortician. A moment later, a long needle was inserted deeply into the base of my neck.

Dreams of aliens and alien faces are now a regular theme. I vividly recall a blond haired girl... very human-like, with the exception of her huge green eyes... of a horse staring at me that morphed into an alien face... of aliens showing me how to hack cell phones with weird magnets and then the next day discovering my own cell phone had gone haywire.

In these "dream-like" encounters "they" have shown me just how extensively we have damaged our oceans with pollution and revealed horrific mutations in the seas. This was before the Gulf of Mexico incident.

I eventually sought the help of a hypnotherapist. I went back to the night I saw a gigantic owl in the Kroger parking lot... it didn't belong there... it was out of place. Its enormous eyes locked onto me and stared into my being.

The next thing I knew, I was observing naked people laying on tables. An alien once again inserted a needle into the back of my neck. I believe it was implanting a monitoring device. I also noticed a triangle insignia.

But the most poignant aspect of this whole matter is the reoccurring "dreams" about the children. Anyone who knows about this subject, knows exactly what I am talking about. I am talking about the innocents... they are a part of us, and they are out there... waiting for us all to wake up.

The Innocents   Monica Geyer

Torn - I have had anomalous occurrences all my life, but I did not connect them to the UFO phenomenon until I was in my thirties. It was then that I read a magazine article listing indicators of "alien abduction" and realized that many of my experiences fit the patterns.

I have an open mind about the cause of the events; I believe that it could be "aliens" from another planet, or perhaps something else that we cannot even comprehend.

When I was eighteen, I was looking through one of my mother's magazines and was moved by a photograph of a baby. When I saw it, I had a strange and profound feeling that I had momentarily "connected" with something important. The image of the baby affected me so powerfully that I rendered a copy of it. As I drew the baby's eyes, I was compelled to make them even more prominent and dark.

Later in life, I had two children who died at premature births. The doctors were never sure of the cause, but it appears that during both pregnancies the placenta became infected, which may have had something to do with it.

When I was thirty-two years old, I had a sudden memory. I was in a bare, room with curved walls that resembled white-washed concrete. I was lying on a single bed or table. A row of about five odd, pale, thin children in white gowns, were standing near my side looking at me. I could not tell if they were boys or girls. They were different heights and appeared to range from about three to six years in age. However, when I focused on one of them who appeared to be biologically about three years old, I "sensed" it was older.

All the children had wispy light brown hair, big eyes, high cheekbones, and little pointed chins. I had an unfathomable sense that they were my children, but I did not know them. Since then, I have continued to have this puzzling remembrance periodically.

Usually, I did not keep my drawings for long, but I did store the one I drew of the baby when I was eighteen. In my thirties, I uncovered it... and even after all those years, the drawing - especially the baby's eyes - disturbed me deeply. On one level, I felt it represented the death of my two children. On another level, I felt it represented a feeling that I was not supposed to have children in my "normal" life, although I simultaneously had memories of children that seemed like my own. I tore up the original drawing but then found myself trying to put the pieces back together. I now see how this drawing is a perfect metaphor for the struggle I have had with motherhood.

I have been on a long journey to understand my experiences. At the present time I am still not sure what is behind them. Are my perceptions of them real, or does the phenomenon tailor itself to me and present me with experiences that I can understand with my own cultural constructs and human consciousness? I do not know. Whatever it is, I feel that for some reason it has been intimately involved in many peoples' lives, including my own. I imagine that it will continue this way.

Torn

Emma Woods

Making A Point - The bright light is shining in the bedroom window... they are coming to take me again.

I'm floating up to a ramp, I'm really cold. I'm in a dark hallway on the ship with an alien. He's not a little gray or a mantis. The grays' faces are flat and smooth. His face is more human, more angular with sharper features and he is taller. He's half human and half alien. He's a hybrid. He is my son.

We are in a dark passageway with a glowing blue light running along the right side of a long, curved hallway. On the left is a big glass wall. We are walking towards two doors that open from the middle. There is a bright light coming through windows in the doors.

My son stops and points to a compartment in the glass wall. The section lights up and it looks like there is a baby inside! It almost looks like a doll, with a fat, little, wrinkly body. It's in a little aquarium.

It is a baby... an unborn baby! He is pretty well developed; a fat, little, pink baby that looks like a plastic doll. He is buoyant, facing me, his little arms floating in the fluid. Oooh! There is an umbilical cord attached to his stomach and floating up around his neck. He has a snorkel on his face, with a clear plastic tube stuck in his mouth. It has black lines in it, like a gauge for whatever fluid is in the tube.

There are a bunch of these "aquariums" stacked on top of each other. Between each unit are two or three inches of black metal framework. They are all joined together. There has to be fifteen or twenty across, seven up and down. There must be hundreds of babies here!

The aquariums are all dark except for the one that he lit up. As he continues pointing to the aquarium that is illuminated, he tells me this baby is from me also. He doesn't speak words, he communicates mind to mind.

I ask him the meaning of all of this. He says I am part of a bigger plan, and that it is a very noble cause, and that I should be very proud.

I am part of a program, not to save the human race, but to assimilate and incorporate the human race. He tells me they're incorporating us to a higher evolutionary step because we're not only destroying the planet but ourselves as well.

They are making babies with my sperm because the Earth's ecosystems are nearing collapse.

My babies are half human and half alien. I would like to see them again someday. It is a wonderful feeling to know I have offspring up there, but it is so lonely to be separated from them and never be able to see them or love them. I find myself crying uncontrollable tears when I realize just how much I miss them.

Making A Point   Chuck Chroma

The Great Room - As my hybrid son walked me down the hallway with the blue neon light, past the incubators which housed the hybrid fetuses, we approached a double door with a bright white light streaming in the window.

Still naked, I was led through the doors and onto a small circular balcony made of white shiny metal. I gripped my hand firmly around the handrail to steady myself as I stood in shock looking out into a room the size of a school gymnasium covered by a huge domed ceiling.

Above me was the most incredible round window looking out into deep space. It had a thick frame of electric neon lights with every color of the spectrum swirling and pulsating within it. It seemed to buzz with energy.

A translucent gauzy veil poured down from the ring like the Aurora Borealis, and then electrified the room as it swirled around the tables on the floor. Below me were hundreds of tables with women laying on them. They were all lying on their backs, paralyzed, and having babies removed from their bodies by the aliens.

Small grays were standing on little triangular steps, while large praying mantis aliens were performing the procedures and seemed to be the "doctors."

They were removing the hybrid babies from the females then putting them into the glass aquariums. There were tables with assortments of shiny instruments, and banks of computers under the big rectangular windows.

I made eye contact with one of the tall aliens. He seemed to disapprove of my presence.

Just about then my hybrid son touched my elbow, and I became aroused. He led me back through the doors and into the hallway. I was then escorted into a small room on the left... I soon realized it was "the sperm collection room."

The Great Room   Chuck Chroma

The Alien Hunter - Most investigators are not abductees. The best information they have comes to them second hand or through hearsay. I am an investigator and an abductee.

The year is 1952. I am between 3½ and 4 years old and lying in my bed... I don't know what is happening or why, but I feel the paralysis hit me as "IT" just stands there, totally emotionless and motionless with those large black, doll-like, shark-like eyes looking at me. I fight the paralysis as the entity comes closer, bringing his face right up to mine. I squirm my tiny frame as hard as I can to get away. Immense fear overwhelms me as he moves toward me...

...I have told you my bizarre, unsettling story as a very young boy. This is the invisible evidence. I ended-up after the second night of events with a scoop mark and a story I was supposed to buy into. "Where's the blood? Where is the scab? Why a smooth dipped scar?"

The Hunted Becomes the Hunter - I had to find out what happened to me and why. I hunt them that hunted me and later hunted my son and may now hunt you or your daughter or son. My investigations began in earnest in the late 1960s, but were limited to business and other casual contacts during my time at NMSU where I was a student and Martial Arts Instructor. This low profile approach was especially true while in the CIA where I was in Covert Training and Black Ops, and in the

Army as a Senior Military Police Officer. "They" [the alien] function like an Intelligence community. I have been touting this idea since 1972. By the time I entered the CIA in covert ops in 1968, I knew quite a bit about how the entity operated and not much about the CIA. I was shocked at how similarly the CIA and the alien operated!

I doubt science will save us nor will our military although I have more confidence in our military than our Intelligence Agencies. I have much less confidence in those who "keep the secrets" about the alleged alien intelligence... secrets of "something" that takes your children in the night and harbors good will as it steals the innocence from them..."something" that robs the hearts and body parts of women and men, then leave them with a profound feeling of being utterly alone.

My Prejudice is that Contact is Real - I have to state: My prejudice is that contact is real. The jury is still out on who the alien is. I am very interested in contact. I am also interested in why these "aliens" only want contact when they can control the entire sequence. I would like to see contact on Alien Hunter terms.

Wearing the Abductee Hat Can be a Difficult One - the abductee hat is a hat of high integrity for me (and for others also). To some folks, this looks like a dunce hat. It is a difficult hat to wear. It forever brands you for being a witness to something others have not seen, for hearing things others were not privy to, for feeling things and telling the truth of your heart about the matter. It is a hat I gladly share with others, though my own events ended long ago at age 17. I didn't volunteer for this one. I didn't choose it. It chose me. I will not lie and say, "The events did not happen."

The Alien Hunter   Art by Kim Carlsberg   Derrell Sims

Beyond My Wildest Dreams - I couldn't resist a cameo...

'One ordinary night, in the middle of an ordinary life, I had an extraordinary "dream." I fell through "the Looking Glass" into another world. However, unlike the children's fable, the place was real, and even more bizarre than the realm visited by Alice.

It has been over twenty years since my first visit to that alien place, and since then I have come to understand many things about it and its inhabitants. But with each new understanding, more questions are raised than answered. Modern philosopher Werner Erhard proposed that instead of constantly seeking answers, we must learn to live in the question of what it means to be human. Whitley Strieber, a fellow abductee, prescribed in his book, Transformation, that we must learn to live at a high level of uncertainty.

They have both described the new neighborhood in which the aliens have compelled my mind to dwell.

I have traced the edge of another reality. I am but a mere child on that foreign shore; tugged by the current of curiosity, while dreading the cold of an uncharted abyss.'

...taken from the introduction of my book "Beyond My Wildest Dreams", subtitled - diary of a UFO abductee, Illustrated by Darryl Anka.

Art by Darryl Anka

In The Blink Of An Eye -

My eyes followed the gentle curves of the tiny face
in hopes of capturing the time we'd lost.
Dim memories of stolen moments in this place,
wondering if she knew what her birth had cost.

My arms, a sanctuary for this perfect child of one.
Time's arrow flew swiftly toward a future scene.
My seed had blossomed like a flower in the sun.
In a single moment, she held the promise of thirteen.
Slender and delicate; a porcelain rose, glowing
with bright innocence, she watched with my own green
eyes. I desired to love her without showing
how much I missed having her in my life.

The fire of her intellect surpassed her emotional air,
which my maternal presence breathed into her spirit.
But she saw the heartbreak it causes to care,
imagined her own future child, and chose to fear it.

I gently reached out to sooth her fears
and reassured her from the depths of my soul
that even a moment with her was worth all the tears,
and for her, like me, the love of a child
would make her whole.

On The Blink Of An Eye   Art by Darryl Anka   Kim Carlsberg

Laboratory - The progenitors follow certain family bloodlines. Speculating on the subject, I'd say that if I was to analyze and study aspects of a civilization, I would do it within the framework of a laboratory.

There would be certain progenies considered the control group and then there would be others considered the ongoing experimental group.

Another scenario would be that a gene line of three or four generations makes for an excellent study. This would tie in with a third scenario...

Question: How would you figure out how something worked if you were extremely curious about its fascinating yet puzzling characteristics or behavior?

Answer: You disassemble it and reassemble it from scratch. In this piece you – the viewer – are on the examination table while three of them are studying you from head to foot. You are being poked and prodded as your mind is locked in an atmospheric, dreamy, drug-induced state of helplessness.

Weeks later, you will develop a nervous, unsettling fear of everyday objects and you won't understand why. Only that you have somehow seen it before – somewhere – in a different place and time.

Pens, knives, scissors, flashlights, forks, screw-drivers. Such things would trigger the images locked away in your subconscious mind.

Adapted from artist's original Visual Arts tutorial lecture - SUNY College at Old Westbury, NY; Spring 1993.

Laboratory

Rick Smith

Dark ET - Early one morning in 1990 I discovered a circle of six needle marks on my inner left forearm.

Later that evening I attended a monthly UFO meeting with about 200 participants. People were welcome to share their experiences within the last month. A night security guard I'd never met, stood and explained how he had seen a small luminous ball zip around a particular house the previous evening.

It was my house, but I didn't say anything. I called Bud Hopkins and he connected me with a local hypnotherapist.

Once I was under, I described being taken out through the closed window and into a ship. The beings that took me were about five feet tall with slightly larger heads than ours, no hair, eyes about half again as large as ours and very blue. They were super friendly and always smiling.

Once aboard I interacted with a very dark ET, about six feet tall. He had me sit in a chair and he put a helmet on my head (to balance the brain waves). I was allowed to view, in detail, a previous life on another planet. This was a place where Nature was studied and experienced to its fullest. One of their

beliefs was after death our connection with nature goes with us, not the study of science. The symbols on the neck band describe how reality is changed through thought and light.

After that experience I was seated opposite the dark ET as he downloaded some condensed packets of information from his third eye area to mine. They had the appearance of tumbling vitamin pill and triangle shapes. These were to be opened later in life. I have since experienced a few of them.

During the question and answer period with him he suddenly looked very surprised. I was being taken from the abduction by another group of ETs from the Andromeda star system.

I have always felt close to this group. A teacher took me to the edge of a cliff. Beyond it was only space and stars. He explained to me what the fabric of the universe was, how it looked and moved.

Then he explained how to create anything using creativity, will power, and sending a thought form through the back of your palm, out onto the pattern. The pattern will morph to your desire, it expands or contracts in a variety of sizes and dimensions.

Dark ET   Danion Kelly

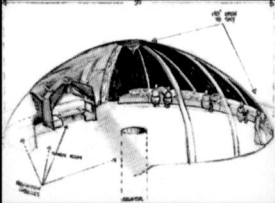

Inside The Ship - I got to see what the temples of learning look like. After that I was sent back to the dark ET. He was astounded that I had been taken. He's very advanced and explained how he didn't need a ship to travel, but needed one to take me.

I then got a tour of the ship. The sketch shows the main level. I was also shown the propulsion system. It appeared to incorporate Tesla looking coils, and crystals.

While painting the dark ET's portrait the following day, it was as if he were watching and correcting me. It is a very accurate representation.

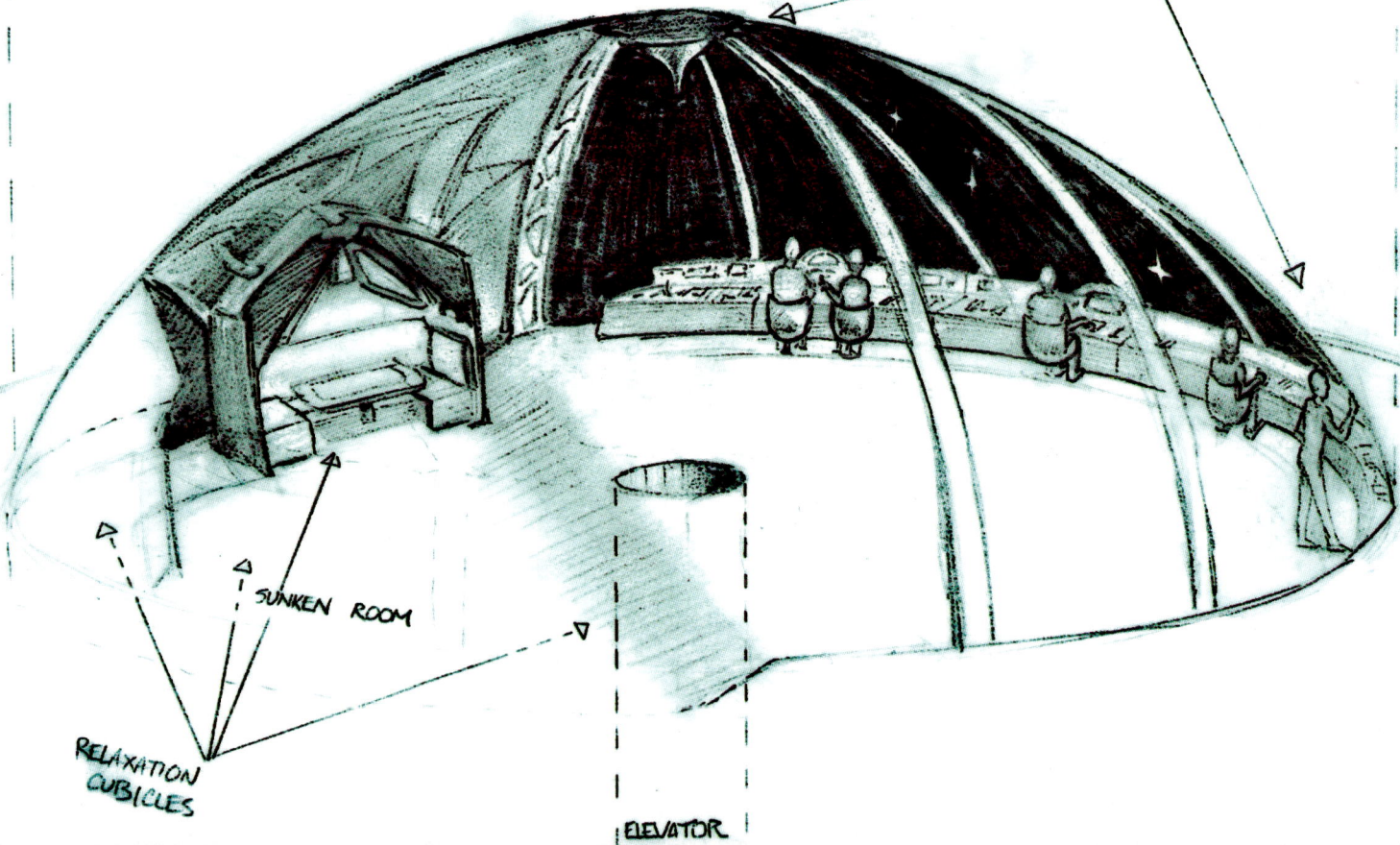

35'

180° OPEN
TO SKY

SUNKEN ROOM

RELAXATION
CUBICLES

ELEVATOR

Taken For A Ride - Wichita Kansas, Fall 2006.
I woke-up, standing in a dimly lit public rest facility. It had the feel of being out in the country; like one you might find at a state park - not one off a major highway. There was another person with me, but I couldn't see her well enough to determine her identity. She stayed behind me and off to my left side. This seems to happen in most of my encounters... I do wonder if she is my guardian angel?

I was leaning against the sink - exhausted, in somewhat of a panic, and wanting a Xanax! A sickening thought "OH NO... not again?" flashed through me! My mystery partner and I then walked out into the parking lot and stood next to my usual E.B.E. (extraterrestrial biological entity) wrangler where she had been waiting for us. She is a sickly white/gray color, who is about 5' 8" tall and very thin. I determined her height based on mine, which is 5' 4".

She was fixated on three craft "showing off" in the sky. The round shaped craft, encompassed in colorful lights, playfully bobbed and weaved before shooting down and hovering over the parking lot. Suddenly three (seemingly male) beings appeared in front of us. My wrangler escorted me up to a well-built being (not a gray, not thin) and introduced me by saying only "This is Angela." She said it in a merry, light-hearted way, as if I were a prize. "Ooo... look what I'm presenting to you!" was the telepathic intonation. The

look I received from him was anything but pleasant; he projected a mean and controlling air. Next I recall being seated inside one of the craft. The area was very compact with a few seats grouped together. The seats were almost exactly like car seats, they were high-backed leather and they swiveled. The color was a light, yellow-beige. I was still with the same beings I had been with on the ground. My wrangler was seated next to me on the right. I asked her where we were going and she responded, "We are going to show you a better way to live." I felt the craft was in motion, but not in a way you would notice when on a plane... there was no heaviness or turbulence.

At the front of the craft was a cavity large enough for one person to access. It was there I took notice of the "Captain." I could see his back through the opening and noticed he was wearing a "pilots" uniform: a white shirt with emblems and stripes on the shoulder area. He had gray, curly hair, but I also quickly realized that he was wearing a wig,... and that HE, was a gray in costume! His head was visibly stressed under the pressure of the obviously "too small" hair-piece. This is where the recall of the night ends.

The next morning, based on my past experiences and the knowledge I have gained, I knew to check myself with the blacklight for evidence on my skin. What I found was unlike any flour trace I had previously returned with: There, on the inside of my right wrist, was a raised, white, egg-shaped mark, approximately three inches long, that seemed to have writing within it. Unfortunately, the only thing I was able to make out were two consecutive "R"s. After recording the mark on my camcorder, I cleaned the area. The mark initially washed away, but as the skin dried it momentarily reemerge before finally, completely disappearing.
- *Art by Garth Perifidian*

Taken For A Ride   Angela Hausinger

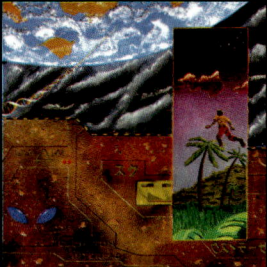

Traveling - The tiny creature hovered. The vibrations in my body pulsed exponentially, converging into a single harmonic overtone. I was propelled along the forest floor as though the electrics in my muscles were animated by a magnetic force.

In ecstasy I glided up, locked in a rhythmic dance with a soul of another kind. My eyes vibrated wildly till our gaze fixed as one and I was aware that this cosmic child had many eyes; babies incubating off-world as many cells morphically entwined, a single group Mind, OBE traveling the corridors of genetic time... And through my gaze peered the shared electrics of all humankind... If only morphically gestating in the womb of spacial-time, I sighed

with a stern love of cosmic impatience. The sigh releasing, unlocking, and I knew: Galaxy, star, sky and tree. Bicameral humanoid - a recurring fractal form in the Universal Mandelbrot Set, born of Gaia to be Her transgalactic holonomic* spores.

Author's note: In Thomas Kuhn's, "The Structure of Scientific Revolutions", Kuhn suggests that a paradigm shift is like the emergence of an entirely new world in place of the old. A number of modern researchers suggest that we may be at the edge of such discoveries. The holonomic paradigm describes relevant developments in physics and biology with profound implications for consciousness and even parapsychology.

ESCHATON
INTERFACE

Traveling

Akira Kawatech

Impressions - It started as a small child. Late evenings, my friends and I would be playing outside and occasionally stop to marvel at strange lights in the sky above.

Since those childhood "sightings", I have had what I thought were "nightmares" of a blinding, white light coming through my bedroom window in the wee hours. I would instinctively try to hide, but I always knew there was no escape.

I shared a room with my older brother, but it wasn't until recently we discussed the odd experiences of our youths. When we simultaneously recalled the peculiar, frightening figure that checked on us during the night, our blood ran cold.

At 18, while laying in bed one night, a high-pitched ringing started in my ears and continued to intensify. A numbing sensation over came my body, starting at my feet, working its way to my head until I was totally paralyzed. A suffocating, crushing weight pressed against my chest... I honestly thought I was dying.

Fear held my eyes clinched saving me from acknowl-edging the obvious presence in the room. Eerily, the mattress suddenly compressed on either side of my pillow as if someone literally stepped onto the bed straddling my head. When I finally cracked my eyes, the darkness obscured any detail within the looming form that towered above me. Suddenly the apparition's hand lunged towards me wrapping around my face, as yet another set of hands grasped my ankles.

After what seemed like hours, instantly everything vanished... the paralysis, the weight on my chest and the mysterious beings. I shot up, screaming frantically, gasping for breath... unnerved and infuriated. These types of experiences continue to this day. I have been aware of the abduction experience for some time, and I am now convinced that I, and several of my family members have been "taken."

Physical evidence has been plentiful. I removed a thin piece of metal from my nostril. I have awakened to unusual scars and abscesses in and around my navel and strange pock or scoop marks on other parts of my body. Once I woke up with a massive hand-print on my arm. I photographed the impression and everyone who sees it has the same opinion... just like this whole phenomenon, it is nothing less than bizarre.

I do not posses full recall of my experiences, but the creature I have painting here is a close representation to one I believe I once saw standing at the foot of my bed, as well as in other flashes of memories in other encounters.

Although my experiences have been frightening, I believe the aliens intentions are not malevolent. They seem to be studying us as we do animals in the wild.

I am hopeful that full disclosure of the alien presence is forthcoming. I believe that it could be the catalyst necessary for mankind to unite, find peace, and evolve to the level required to be openly interactive with these other galactic occupants.

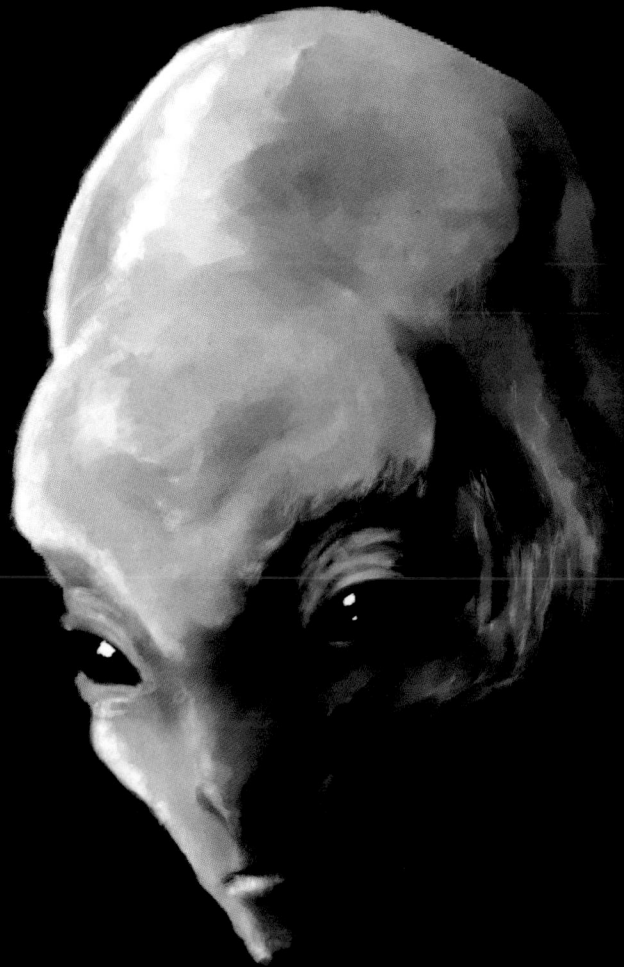

Impressions   Andrew Pearce

161

Abduction Seduction - I remember an evening in particular... which started in the waking state. I had been painting: "The Raven & The Serpent", my personal metaphorical transformation for the integration of the light and the dark, in this never ending spiral of awakenings. I felt the evening was early and decided to impeccably groom - as if for a special honeymoon. Why that thought crossed my mind in that way I had never explored until I began "the remembering."

Every inch of myself was glowing, oiled, glittered, scented & decorated in my usual Goddess Sacred Dancewear. Nine or so rings including toe-ring, six earrings, headbands, swords, candles, and the most magical sacred music, including veils. I was alone. Or so I thought.

Just as if on stage, I "showed up" as the real thing. "Layering in" essence of Goddess. I danced and breathed myself into oneness with myself and "all that is." Suddenly, there was an acute awareness that it was being experienced and recorded from an "outside" point of aesthetics. Who and what is this "outside" space invasion phenomena? This sexuality is experienced as a total mind body soul explosion which can only be described as a continuing river of velvet purple aqua water in slow motion...

like Dolphins flowing together and apart, always in synchronicity, yet independently creating. That was the ideal. Feeling this etherial 'partner' as physically present ! ~ I experience sheer delight for our timeless connection and full recognition of the sacredness and timelessness of this ancient/future union. I was the song played. I was the dance... the place where you say "yessss" - and it is golden and divine.

The ecstasy is something no words can harness. Find me in the realm of the Sacred... the honor of "conscious" experience.

My Personal Side: This painting felt unique to me, as I knew this "experience" wasn't a dream. I had evidence of interaction and a heightened awareness of Divine Tantra, and clear telepathy that came up while in my higher conscious state after this episode. I became aware of a very insightful and 'new to me until then' aspect of the abduction phenomenon. I became aware of the inter-dimensionality aspect and how layered it was. All of a sudden I understood "Alice in Wonderland" for the first time. All wasn't as it seemed to be!

The layered dimensionality of the experience was frightening at first but more for the disorientation than the perception. At that time I had also just read the Jane Roberts: "OverSoul 7" trilogy and it brought the hologram awareness right into present time! The flexibility of it all!

Then there was left for me the Great Mystery and a yearning for what I could not remember from this interaction that haunted me until the next encounter... which is another story in my upcoming revelations of these on-going series of paranormal unfoldings to date. So be it -

Abduction Seduction   Raven De Lumiere

Alien Tourists In Hell - Aliens and UFOs have fascinated me ever since I was a child. I've never seen a UFO myself, but I may well have had some sort of abduction encounter.

From age 16 to 23, I would have strange episodes when I was sleeping. I would wake up and there would be a glowing, blue field around me, the field would always be reflected on the walls and ceiling, like the reflection of water.

I wouldn't be able to move or speak and I felt someone was standing just outside my vision. Eventually, I'd be able to move again - just a finger to start with, and the light would slowly fade.

If someone had been sleeping with me, they'd still be there, fast asleep. If I woke them, they would always say they hadn't seen or heard anything unusual.

I'm still not quite sure what was going on. I've read up on experiences like it, experiences similar to night terrors, but somehow I think it was more than that. I'm in my forties now and I haven't had anything like that happen since.

Alien Tourists On Hell

Cliff Hare

Mind Scan - My name is Sanni Ceeto. I am an abductee. My abductions started at age two when I was circled by a tiny, silver ball in our yard.

I am also a hybrid... I am a "Zeta-Terran." My father is a thin "gray." My mother, who has passed, was also an abductee. She was told by the night-time visitors who took her, she would have an unusual child... me.

Both my Earth parents died when I was six. I was told their car was hit by a train. I was moved and kept in a glass cell on what I am sure was a military installation but told was a hospital. I was submitted to a battery of tests, physical and mental, throughout my early years.

One experience I remember in particular was being visited by a man with a goatee, wearing a dark blue uniform with lots of stripes, stars, etc. He handed me a piece paper and a pencil, and said... "So this is it"... it referring to me. "It must never find out about itself." He tells me to draw the sky.

Most children would draw a single sun... he looks at my drawing, I had drawn two suns. I was eventually moved into a foster home were my foster parents were also told that I must be shielded from learning

about "myself." "Myself" meaning my ancestry, and my other "home" Zeta Reticulum... "a binary star system."

Everything in my life was censored, but the single biggest area was UFOs. In this foster home, my "progress" was monitored once a month by a "doctor" who would visit, study me, ask questions, take notes, and leave.

"My people" have visited and monitored me over the course of my life. The "tall-one, Khineo", told me over and over again "we are as one, you are of us." I become one of them when I am with them.

This particular event occurred as an adult. The observing alien seemed to want to share or intrude into a private "mind scan" I was engaged in with another alien. This actually irritated me, and I believe it irritated my companion as well. I underwent a full mind scan procedure, our foreheads touching, we "locked" into each others eyes... my thoughts were examined and there was exchange of information. The information exchanged dealt with new life-style changes I chose, such as becoming vegetarian, using herbal medicines, healing with crystals, channeling, and my deep interest into what humans call "New Age."

Mind Scan

Sanni Ceto

The Space Aliens - Finland, Thursday, Feb 26, 2004 in the countryside at my small farm. There was snow and little sunshine, temp around -10 Celsius. I was watching TV avoiding a strange new phenomena - "small voices" in my head - that had begun earlier that evening.

Suddenly, a small, bright white UFO, the size of a football, pass through my yard at remarkable speed. I thought I was hallucinating when an inner voice said "go to sleep now." I was amazed, but felt no fear. I went to my bedroom... the radio was playing. "She" asked me if it disturbed me... I turned it off, and reclined on my back. I was very excited.

The voice instructed me to close my eyes. As I did, a blast of light hit me and I could no longer open them. I asked how this could be happening and the reply was "don't worry about that now." I spoke words, she answered telepathically.

Surprizingly, a warm, relaxing sensation spread through me as (now more than one voice) made small talk. Somebody suggested that I be teleported somewhere, another voice denied it. I asked them where they would have taken me. The answer: Three Sister Mountains, Oregon, USA.

I also felt the presence of "others" and asked if a UFO was nearby. I heard laughter, and a response, "just a small one, at the top of your apartment." They quietly proceeded to examine my entire body. "The abductors" were neutral for a time, but the following nights I had terrible dreams as they tested my psyche with nightmare scenarios. The next few days, I tried to continue with my daily routines, blocking out their influence, but the aliens would manipulate my mind at their will.

At one point, I was invited to join them, to travel to the States and be introduced to their secrets. They pronounced themselves as part of the military of the U.S. as "humans in alien form." I would give up my body to inhabit an alien one. I sensed one of their secrets had to do with mind control on a massive scale. I was not interested whatsoever. Due to limited space I cannot go into details, but I can assure you the interactions were bizarre beyond comprehension.

In short, this is a taste of my three day "abduction." It left me with so many unanswered questions. As a physician, having worked in mental hospitals as a resident doctor, I knew I wasn't mad. But following these strange few days, I became so depressed I was eventually forced to seek therapy.

The abduction experience was an abrupt and complete transition from the known to the unknown. The impossible was suddenly palpable. One is stripped naked in front of the aliens and realizes our assumptions about reality are nieve.

A huge segment of our world's population is not aware of the reality of this phenomenon. Military secrets are highly guarded. FOIA requests will not get this information release to the public. It is up to us to do what we can, and the truth "IS" breaking out.
Farm image by Jacki Mroczkowski@imagekind.com

The Space Aliens   Art by Kim Carlsberg   Usko Ahonen

The Fear Egg - My most memorable encounter with this being occurred when I was about seven or eight years old.

The encounter started the same way that they all do. I awoke in my bed completely paralyzed. There was a bright blue light in my room and I could just see the tops of three bald white heads moving around my bed.

I felt myself being lifted from my bed and then right through my closed bedroom window. At the time I was completely unaware of my previous encounters, so this was all new to me and therefore completely terrifying.

I don't remember exactly what happened next, but I know that at some point I was taken into a large round room with a large blank screen upon the wall. There were several old style school benches and chairs all facing the screen as if it were a black board in a classroom. There were no other children, but there was a single being at the front of the class.

I was brought before this being by two grays. I was semi-paralyzed, cold, naked and completely terrified.

The being in front of me was dressed in brown monks robes and had an odd looking wooden staff, with a silver globe at the top.

He had an incredibly wrinkly face, wide set black eyes, a flat ape-like nose and a huge heart shaped head with bony ridges running across the groove that separated the two hemispheres of his huge ugly head.

Home World - I found his appearance completely monstrous and my terror multiplied up to the point that I felt as if I was falling.

The being then started to shout at me. The sounds were in my head, but he even went as far to mouth the words as they appeared in my head. The effect was really weird and slightly comical, as I could also hear the odd ape-like grunts that he was making as he mimicked real speech.

He was saying things like how dare I be afraid of them. I was being given an opportunity that few humans got and I was treating them like monsters. He then started to call me things like stupid monkey, and stinking empty-headed fool.

His insults became more and more childish and eventually my fear transformed into anger, so I reacted. I started to throw a few insults back at him, becoming angrier and more sure of myself

as I did so. Suddenly he threw his head back and started to laugh, not just in my head, but for real, big heart-felt belly laughs.

He then congratulated me on defeating my fear and told me that fear is like an egg. It is impossible to break the egg from the outside without harming the chick, but given the correct environment it is possible to encourage the chick to break the egg for itself.

Ever since this encounter I felt more comfortable with this being than any other that I have met since. He always treated me as an equal and has been a powerful teacher throughout my life. He has shown me things that I think no other being would, including his homeworld.

He fed my curiosity with knowledge and I am grateful for his presence during my early childhood encounters.

Home World

Stephen Martin

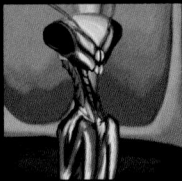

Weirdest Year - 1996 was an eventful year for me; I had a complete out-of-body experience that started with an intense spinning sensation and within weeks I had two close encounters with nonterrestrial beings. My memories of the close encounter events were fractured so I sought the help of a hypnotic regressionist. Earlier that year I had already called on the help of regressionist and Ufologist, Eric Morris after finding his number in a local UFO magazine.

This time I asked him to help me uncover memories from my recent spate of encounters. The first was a fairly uneventful encounter with a single entity that I will bypass to get to the much more interesting encounter that is the subject of my drawing.

My pre-hypnosis memories of my second encounter were of simply waking in the early hours of the morning to find two non-human entities standing near my head. Under hypnosis, I again remembered awaking to find the same two beings in the same position. The first was a slender gray wearing a hooded black and grey cloak. The second figure was a wrinkled, brown entity who was wearing a brown hooded robe.

The next thing I remembered was being lifted from my bed, through the window and up into a vast circular craft. I could not recall entering the craft, but I did remember being naked and led through a large, cold room full of many different entities. There were many grays, but also, large insect-like beings and two huge lizard like creatures that were about nine feet tall. I saw more wrinkled brown fellows like the one that I had seen next to my bed and I remember seeing a single pale entity with milky white eyes.

I was laid upon a low metal bench next to a wall under a huge screen. From this position I saw just how vast the room I was in was. The walls were a dull metallic grey color and arched with long curving pillars all of which were topped with large oval shaped white lights. As I lay on the bench the little pale fellow stood by my head as two grays moved a sheet of metal over my lower body.

Momentarily I began to feel myself spinning, just as I had one month before when I had my out-of-body experience. I realized that the visitors were trying to initiate another out-of-body experience. The thought of being pulled from my body in front of all these creatures terrified me, so I tried to bury my consciousness into the back of my mind and hold on with all my might.

The spinning became more and more intense and while I was in my regression and I actually became concerned that I was about to leave my body right in front of Eric. Luckily, at that moment, there was a knock at my door and I instantly jarred out of my hypnotic state. The sensation was nauseating, as if being in a broken elevator that suddenly dropped to the bottom floor. I have never managed to recall what happen during that encounter, whether I left my body or not, or what happened there after. I have considered trying, but I have never found to courage to relive the intensity of that experience.

Weirdest Year   Stephen Martin

of planets. When both stars/suns are high in the sky outdoor travel is forbidden, simply because the solar radiation reaches lethal levels.

The whole experience was like watching a documentary. I was told that during the hot season that two of these giant reptile people must make a journey across the desert, due to some emergency that I didn't understand.

I saw a ramp descend in the desert floor and after a couple of seconds two of these beings came racing out. When these guys run they do so on all fours. I saw these guys sprinting across the desert trying to reach the nearest town, before the radiation killed them. The images end with them intercepting a large long wheeled vehicle that was trundling across their path. They finished their journey, in relative comfort.

It all seems rather complex and protracted for a dream doesn't it? I believe that I observed these images as a holographic projection. I have experienced this during past encounters, I was once shown a possible dark future for our planet. I was shown my own home town after some kind of terrible event. The air was choked with smoke and the low lands were flooded.

At the start of the experience I was watching the events on a television screen that seemed to hover just near to a wall. As I watched, the screen expanded and eventually surrounded me. One minute I was watching the images and the next I was in them. I could even smell the smoke and feel the wind on my face. I suspect the images where I saw these giant lizard/reptilian beings were something similar.

Lizard Bloke - I saw two of these giant fellows during the same encounter that I saw the praying mantis like beings. At about nine feet tall these guys kinda stuck out from the crowd.

About five years after my encounter, I woke up one morning with some weird unconnected memories in my head. For a long while I was convinced that I had had some kind of lucid dream, although the dream had details that I hadn't seen during my encounter. I remembered seeing these beings on their homeworld. During all this I knew that there was someone behind me narrating everything that I was seeing.

I saw a huge brown dusty desert. I was told that this world spins on its side, so this area of the planet always faces the planet's main star. This desert is locked in perpetual day, so its inhabitants have learned to build their towns and cities underground. I saw buildings actually carved into the side of canyon walls and cliff sides.

I was told that this planet has a long and strange seasonal calendar, due to a second smaller star that orbits their main star on the very edge of their system

Lizard Bloke

Stephen Martin

Detour - In 1990, two male friends and I were abducted and experienced missing time while driving on a mountain road. We thought we saw a UFO in the sky and argued about it when suddenly there were three blinding flashes of light and the driver stopped the car. The driver later said a force took over the car.

A four-foot tall alien being was standing in the road and three more beings appeared, two opened the front passenger door and another opened the driver's and both friends were removed. The car's back hatch door opened and suddenly I found myself floating up and out the back door with two beings, and was set down standing in the road.

To the left where the roadside dropped away hovered a UFO with a ramp coming down. One friend walked onto the craft ahead and the other was escorted by two beings. I was taken by the hand and led up the ramp. On board our clothes were removed and we stood naked in front of each other.

We were taken to two examination tables and a reclining chair. One friend got onto a table, the other into the chair, and I onto a table. The beings proceeded to remove ova from me. My friend on the table was positioned to watch, but resisted, turning his head away. Later he said they tried to extract sperm, but discovered he had a vasectomy. My friend in the chair had headphones placed over his ears.After my procedure, I was taken to another room and a large group of beings circled me. A black bag was pulled over my head and tightened below my chest and pinning my arms. It was completely dark inside and I was pushed sideways and started to fall over. I couldn't move my feet and realized if I fell I wouldn't be able to break my fall. I was then pushed over repeatedly, each time in a different direction. I felt like one of those blow-up clown punching toys and was screaming and begging them to stop. Eventually I realized they were not going to drop me and the less I resisted, the less they pushed me.

After stopping they removed the bag and offered a weak apology saying I had to learn I could trust them. I was angry as they took me back into the other room. One friend later said he had no idea where I had been, but that I looked really pissed off when they brought me back.

While I was out of the room, the beings gave one friend some technical information and showed him how to build a UFO detection device. He eventually made these devices and always claimed this is where he got the idea. We were taken to where our clothes were and instructed to get dressed on our own. The beings took us back to the car and we were positioned like moving a manikin, back to exactly how we were just before stopping.

We then drove away as if nothing happened. Suddenly the car radio wasn't working and making strange buzzing sounds, yet we had no memory of what had just occurred. One moment we were arguing about a sighting and the next the radio is going haywire. We didn't realize until later, that two hours were gone, and there had been a detour.

Detour   Art by Dean Wolfe   Melinda Leslie

Dream Encounter Art

UFO Abducted? - What I remember best about this "dream" is that I'm on my stomach, face down on some kind of metal table with my head upright - yet tilted to the left somehow.

I'm in a large, circular vehicle that I immediately realize is an extraterrestrial craft. I don't remember getting a good look at the area or objects around me even though it was well-lit, it seems they were in a kind of haze or blurred background. Interestingly, I wear glasses most of the time because I'm near-sighted.

Behind me, are "they" who are responsible for paralyzing me and placing me in this prone position. Although I can't see them because they are behind me, I know the ETs are the infamous "grays."

Somehow, I also know it is not the first time I've been in this bizarre situation with them, and I believe they previously found me to be amusingly annoying.

Perhaps it has something to do with the fact that I am filled more with anger and resentment than fear toward them. I recall I inexplicably sensed there were also regular people observing and even assisting the ETs, presumably government or military individuals. I cannot speak, but I am aware they can hear my thoughts and so I angrily say, "Why do you have to paralyze me like this, huh? Why don't you try doing this to me without paralyzing me, or are you afraid of what I might do? Do you know what I can do to you with my hands and feet?" - referring to my martial art background.

They replied, "Do you know what we can do to your hands and feet?" I suddenly see the image of a darkened room with a kind of spotlight shining down on a very large rectangular tank of transparent, greenish liquid. I then see my severed arms and legs splash into it, causing bubbles as they quickly sink toward the bottom.

I was quite taken aback, to say the least. I definitely didn't expect such a vivid reply from them. Before I could formulate my "And yet you still have to paralyze me in order to do this to me, you cowards!" I awoke.

That whopper of a reply left me feeling as if I had an actual conversation with smug beings apparently suffering from a massive superiority complex.

This unusual "dream" occurred in the '90s while I was living on the north side of Chicago.

UFO Abducted?   John Pagan

Dream Within A Dream - This was many years ago, perhaps during the late 1970s. I was about eleven years old living with my parents, two younger brothers and baby sister, in the north side of Chicago.

One of my brothers told me the previous night he "dreamt" of seeing a UFO in our backyard, not too far from the wooden porch of the first floor.

He was able to describe it in detail and even drew a simple sketch of it; saucer-shaped, lights around its edge, and a red light on top of it.

This was so strange to us because I too, had a bizarre "dream" that same night which I immediately shared.

I was in my dimly lit bedroom on my bed, when I suddenly felt a strange, unsettling sensation, as if I knew something or someone was coming. I quickly yanked the blankets over my head and hid under them in a fetal position.

An intensely bright light suddenly appeared to come from the wall in front of my bed. I knew a familiar

"they" were coming for me and closed my eyes to feign sleeping.

As if dreaming in the dream, I could also see multiple dark human-like silhouettes emerging from the brilliant light that was once the wall, even though I was curled up in fear with my eyes closed under the blanket.

I felt completely vulnerable and exposed, I realized that they knew I was awake... they always seemed to know exactly what I was thinking.

For a moment, I thought I could hear them in my head saying something to the fact that they wanted me to stop pretending I was sleeping. When I realized they were about to remove the blankets, I woke up.

We never mentioned that night to our parents, we knew they would tell us it was only a dream and send us on our way.

Could it have been simply a night of over-active imaginings of two kids? Possibly. But it sure made for one heck of a coincidence.

Dream Within A Dream   John Pagan

Hunt For Reticulans - Growing up in America, in a house where creativity and imagination were nurtured and encouraged, I developed a mind open to all possibilities.

The idea of little green or gray men, and space invaders, was never a concept that I ignored or rejected, and when I began to hear incidences of crop circles, animal mutilations and spaceship or alien sightings, my curiosity became overwhelming.

But it wasn't until I became active on the web, that I found vast amounts of reading material available about these things, and other people who were interested in the same areas.

I personally saw two questionable happenings in the night sky. A huge flash of streaking light over L.A. was explained away by the government, on the news. I had seen it soaring over Hollywood, as thousands of others had and I could not be dissuaded from what I had seen. On another occasion, I saw a hovering disc shoot away into the inky sky, and once again there was an announcement on the news saying there was testing. I had seen these things myself; was my imagination running on overload?

Then, I came upon a website called "The Black Vault", which supposedly housed information that had been from secret files released under The Freedom of Information Act, which told of the discovery and cover-up of many sightings, and alien abductions.

One night soon after, I dreamed of an alien who visited me in my bedroom, who asked permission to use my younger face and body to visit the Earth and to capture aliens who lived secretly underground. When I woke up, I began to draw.

That is how my Supremextreme character came into being." "Hunt For Reticuluns" is a portrait of that character, overpowering an evil alien, and sending him back to the skies.

Hunt For Reticulans

Susaye Greene

The Man Who Volunteered - The next four images are excerpts from a website I created and illustrated in an effort to discern the truth about whether or not "a dream" that I experienced several years ago was:

     A. ...just a highly unusual dream or...

     B. ...some sort of alien message telepathically delivered to me during my sleep.

"Garth's Dream" can be viewed in length @... www.freewebs.com/garthsdream. As the dream unfolds, I saw what appeared to be average houses built up in a somewhat woodsy area that seemed to resemble Earth. There was nothing to indicate that this community was anywhere else but here on Earth, until that is... you looked up into the skies!

The next thing I saw were the ships, standard flying saucer-fare. They hovered quietly above the community, day and night. They watch for signs of unauthorized activity. Beams of energy spit down from them now and then whenever they saw movement at night.

The Man Who Volunteered   Garth Perfidian

Whistler - Here's an odd thing... there are other subservient alien races living on the planet.

They are few in number and build no domiciles for themselves. They are dumped infrequently onto the planet surface and live in abandoned human houses.

Pictured here is an alien called a "Whistler." The name refers to the weird mouth on the thing which effects a perpetual 'O' shape as if it were constantly whistling.

Its face also seems to resemble some Native American ceremonial masks.

Whistler

Garth Perfidian

191

The Night Wolves - They are 'genetically enhanced' wolves that prowl the community at night looking for a fresh meal. That meal might be human or a stray alien, it makes no difference to them. Anything that moves is fair game.

Now when I say genetically enhanced, specifically, I mean... THEIR SIZE! These wolves stand SEVEN FEET tall at the shoulder! They are the perfect means of population control at night.

Their enhanced sense of smell, their speed and their vicious joy of hunting make them perfectly suited for the task.

The Night Wolves   Garth Perfidian

Female Alien - These albino-like females are usually kept on the ships (or elsewhere on the same planet or on another planet or space station) and their function seemed to be primarily sexual in nature (or perhaps just as servants who are easy on the eyes).

I am guessing that they are bred to be small in order to be more easily dominated. Taken as a whole though, these alien females seem a rather salty lot. They may be genetically engineered to have insatiable libidos that drive them to serve their masters.

I believe this particular female was deemed overly "troublesome" for some reason and booted off the ship as punishment.

Obviously, these creatures must work on the ships in order to endear themselves to their masters and advance themselves (and/or gain some measure of security). If they are not allowed on the ships, they cannot advance themselves.

So this punishment is severe in most cases, but I got the distinct impression that this female didn't much care one way or the other. Her feisty attitude and belligerence seemed to be the reason why she was exiled.

Female Alien   Garth Pertidian

Shape Shifter - was inspired from my experience of terrible nightmares in my college years. They were haunting embodiments of people I knew who transformed into other worldly beings and demonic entities.

I remember feeling so tired in my classes and just wanting to get some sleep, but every time I tried I had nightmares. This went on for three years and then they suddenly stopped but I never knew why.

This digital artwork is a reminder to me of my mental encounters with shape shifters. It is meant to give the viewer the feeling of a changing human form in a dream like environment.

Shape Shifter   Laura Barbosa

Who Are The Aliens - was created from a dream
that I had about a flying spacecraft hovering above
an old alley.

When I entered the alley I found an old car with a
strange being in the drivers seat. He was attempting
to take a woman hostage, who was in the front
passenger's seat.

Her body was hanging partially out of the car
door and I could see that she was missing a foot,
it seemed to have been severed off. In the dream
when the driver turned around to see me watching
them, I woke up.

I feel like the car represented our government,
the UFO in the sky represented things that are
happening that we have not been told about.

The driver taking the woman and causing her harm
represents the effects that these secrets have on us
as a society.

Who Are The Aliens   Kristi Kennington

Adam - I do recall as a child a vivid dream about two glowing figures entering my bedroom and causing my covers to spin and twist away from me, and my mom coming in and getting upset that my covers were all twisted.

I asked her the next day if she came up to my room and she said she came up there a few times and every time my covers were twisted up. I recall after that having several nightmares about floating in a spinning vortex of forced air, once floating down the stairs from my bedroom. Then I'd wake up terrified.

Never recalled any other aliens, though. But in retrospect it sounds similar to other experiences.

This image supports the "aliens made man" theory.

Adam   Dale Ziemianski

Contact Encounter Art

The Elder - This stunning and wonderful ET is one of the elders of a race of "Beings" I have dealt with since I was born. Her energy is so gentle and compassionate. When you are in the presence of any of these Beings your own energy seems to vibrate at a much higher level... you can feel their love pouring over you.

They used to visit me during the day when I was a little girl, appearing on a platform in the sky, quite low so I could see them very well. I was always afraid that others might see them but they told me that I was the only one who could see them and not to worry.

I'd spend hours talking to these amazing Beings, and felt so homesick while I was with them. I would always ask if I could return with them, but they explained to me that I had a lot of work to do in my life that would help the world understand that they were real, there were many other ET races out there and that they were part of our extended family.

I never had names for my Beings as I could always tell which one was going to make contact simply by their energy signatures. I was extremely lucky later

on to be taken for a visit to their home planet. This complex wasthe most beautiful and incredible place you could ever imagine. Their structures were "grown" I was told and shown part of this process. There were no right angles or sharp edges on anything in the rooms, large buildings were circular, everything flowed. I asked them about this and they told me that long ago they had realized that negative energy can get trapped in sharp right angles so they had gotten rid of that design many thousands of years ago and now used only architecture that completely allowed the energy to be free and not get "stuck" anywhere.

This beautiful Being was with me the whole time during my visit, showing me their medical facilities (which were hardly ever used). The ETs showed me how they heal injuries using the most incredible technology. I was shown how their crafts were made, how the technology works, it was the most magical time of my life. I desperately wanted to stay but again they told me I had work to do here on Earth and one day I would be with them again.

In all my paintings of my wonderful ET Ladies, you will see they "appear" to have make-up on. Well I can assure you it's not make-up, they look like that "naturally." They even allowed me to 'attempt' to remove it using a damp white cloth. I scrubbed, and scrubbed to try to get the lipstick off... nothing. I tried to remove the eye make up... nothing. I felt her eyelashes and they were so long and "lush" but they were totally natural. I always wondered from that day if we here on Earth created make-up to try and look more like them.

The Elder   Karen Lyster

Little Miss Bright Eyes - I call this beautiful Being "Little Miss" for a number of reasons. First she is the youngest of the ET Beings that I have dealt with. The second reason is because she is so mischievous she acts a lot like a teenager does even though she is over 500 years old.

Yes she thinks it's incredibly funny I call her "Little Miss" when she is centuries older than I am! She has a delightful personality and unlike the "older" and more mature ETs, is very animated in the way she moves and gestures (if I had to put an age on her in "our years", I would say it would be "sweet sixteen").

When I visited their planet she buzzed around me so much of the time, wanting to know what I thought of everything and loving my "strange" reactions to what I was shown and told. They all use telepathy when communicating but she also has the most delightful laugh, which she displayed when I tried and tried to remove the "makeup" off two of the other ET ladies. Which as I previously explained I couldn't remove as it is completely natural. I remember her clasping her hands together and putting them in front of her mouth in great anticipation of my reaction as I took the white cloth they had given me so I could see for myself that their "makeup" didn't come off as it "wasn't makeup." She jiggled around with delight and jumped up and down as I scrubbed and scrubbed their faces and got absolutely no where, she thought that was so incredibly funny.

Her eyes are as amazing as the other ET's, but because her face is slightly smaller they seem to stand out more and are the first thing you focus on when you meet her, which is why I also call her "Bright Eyes." She's sweet, funny, so beautiful and her energy is like walking in a soft summer breeze of contentment that just washes over your entire Being and totally relaxes you. Just amazing my "Little Miss."

Little Miss Bright Eyes

Karen Lyster

The Lightworker - She has always been there throughout my life protecting me. Two years ago we had to purchase a new car because our old one had completely done its time. She warned me that not long after we purchased the car we would be in a serious accident and be hit from the back by a white truck. Our survival depended on me choosing the "right car." My poor partner must have shown me well over 50 cars before I felt "yes this is the right one, we can survive this."

Since I knew we would have a bad accident and the car would be totaled, when we purchased the car from Ford Motors we asked the salesman if he had another one just like the one we were purchasing. He told us he did, so we told him "hold it for us." Of course he thought we were completely mad as "why buy another one when you've just brought this new one." His actual words were "you've got to be yanking my chain."

Two weeks later we were driving down the motorway and in front of us was a large white truck. To our absolute horror we watched the metal part of the entire back of the truck come off and start wheeling it's way towards our windscreen. "OH GOD" I thought, "I'VE GOT IT WRONG, IT WAS FROM THE FRONT, NOT THE BACK." I braced for the metal to come through the windscreen and most certainly into my face. To this day I have no answers as to how my partner missed this enormous metal object flying its way towards us but he did. But in doing so we braked so hard that we were "hit from the back by another white truck!"

The only part of our car that wasn't totally destroyed was where the two of us were sitting. The entire back of the car was a complete wreak. I remember getting out of the car and immediately feeling so sorry for the young man (who was only 18 and his first day at work!) who had hit us as he was so upset. I hugged him and said how it wasn't his fault, there was no way he could have missed us. I even said to him (in the heat of the moment), "I'm so glad it was someone as lovely as you that hit us"... heavens he must have thought I was nuts!

Well, you can imagine the look on the Ford dealer's face when we turned up three weeks later after purchasing the car and asking for the "other one he had on hold for us." He just couldn't believe it! My wonderful "Lightworker" had told me exactly what was going to happen, and I owe her my life many times over as she was helped me all throughout my life. I love her, I respect and honor her, she is an amazing Being.

The Lightworker   Karen Lyster

Aqua Eyes The Mystic - This particular Being has the most amazing eyes I've ever seen. They almost seem "fluid" as if you are looking into the wonderful aqua, blue-green, glacial waters of Antarctica. I remember when I first met her, her presence was so strong I lowered my head as I felt I was looking at someone who was extremely respected and very ancient.

It turned out that is exactly what she was. She is thousands of years old and helps me in my teachings to train my mind to manifest items or control my surroundings.

Over the years I have often been taken to what I'd call "classes" where this Being is the instructor (along with two others). One of the exercises she taught me was to levitate a ball - very simple to them, but very difficult for me. The ball would have a piece of string like material attached to it, so it was the string I would pick up and then of course the ball would move up as I took it higher into the air. I would spin the ball around, make it go in all directions - it was easy because of the "string." She would then tell me that I was only imagining the string being there and that there wasn't any... so in essence I was moving the ball around with my mind. Of course the

moment I was told this, the ball dropped to the ground simply because my human brain couldn't imagine how I could move the ball without it being attached to "something." We do a lot of training like this, including walking up a white set of stairs (which weren't actually there - only in the minds eye) then stepping off the top and floating down to the ground. I had to do this one over and over again (I still do) as I would always get nervous half way up those "non-existent" stairs and be afraid I'd fall. But even when I did get those nerves and the stairs would disappear she would always gently hold onto my arm and we would float together to the floor.

As I mentioned, her energy is extremely strong - but the most loving and compassionate energy you could ever imagine. Nothing I did or do is ever "wrong" during my teachings, she was and IS so patient with me. Sometimes I wonder what she really thinks of this strange human girl they seem so interested in. I'm not normally a shy person, but I am when I'm with her. She is so magical and mystical and so much more advanced than I am, that I feel so unworthy of even her attention. She always knows what I'm feeling and tells me I'm so wrong and how well I'm doing.

She showed me once how she creates her environment. She took me outside to one of the courtyards (extremely large between buildings) and as she walked, flowers would spring up along both sides as she passed by. She loves the color violet, so most of the flowers were that color. She's an amazing Being and one I am very honored and humbled to interact with.

Aqua Eyes The Mystic

Karen Lyster

Beyond The Veil - One night I was compelled to attend a gathering. Normally I would never think of attending an industry function as I work hard and late. But for some reason I felt a strong urge to go so I did, against the wishes of my girlfriend. She decided to stay home. We quarreled about it, but I still went.

Upon arriving, the usual notables were there looking to further their projects or careers. Eventually work talk subsided and people slowly mingled into small groups to socialize. As I surveyed the room I took note of three groups, but one in particular. This group seem to be completely captivated by one young, almost perfect looking man, who couldn't have been older than eighteen. I was never able to find out what he was speaking about, but what followed changed me.

As he continued to talk, he turned towards me, and with out pausing, spoke to me in my head. I was shocked and immediately turned away in fear. I was almost afraid to look back in his direction, but I did. He spoke again mentioning something personal. He said that I knew I was going to get in trouble when I got home that evening. I once again put distance between us. I was not able to deal with it right away. I felt so exposed. It was a new feeling. As the evening winded down, I finally got the courage to approach him. I hesitantly asked "That really happened, didn't it?" He said "Of course, your just one of the few who can hear", meaning there are not many... that I was lucky. He stood up and shook my hand. When he did, something astonishing happened, his hand was warm and I felt a jolt - saw

a flash - then he was gone! I was suddenly outside in the cold night looking up at the stars and seeing them differently. I saw a pattern, a rhyme and a reason they are arranged the way they are. But he was gone. I had lost track of time, but I recalled the flash... it came to me suddenly, all the images, all the feelings, the choice we all will have to make soon. I can tell you what I was shown taking place around the globe, but I feel an old Hindu story might work better.

God becomes fascinated at how a pig is so happy living and playing in his own filth, so God decides to become a pig. Eventually God is so captivated with his new life that he forgets that he was ever God and lives quite happily until one day God is called upon by his contemporaries to come back. There is much work to do but he says that he is only a pig and ignores the pleas. So in a last attempt to rescue God, they kill the pig, and God emerges. He says he is sorry he forgot himself and he returns to his former duties. You see Earth is the pig. We are all lost. Its almost here. Everyone will have the same choice soon.

Most of us will be afraid... it takes practice to learn how to maneuver on the other side. The security blanket of our former selves is easy to recreate there, but we need to ascend.

Every soul is like the seed of a dandelion. Each is a valid experiment in evolution and potential. When we are sent across the wind, some find soil and grow, others find rock and water and are lost. Every soul has potential and all have the opportunity to move forward. But some stagnate and will be lost. They will miss the opportunity, and they will go back into the light, while others will become the light.... like the source. What I was shown was hard to watch. They are here. Waiting for the event to come. It is as regular as the seasons. We are lucky.

Beyond The Veil   Art by Peter Nunnery   Gary Purviance

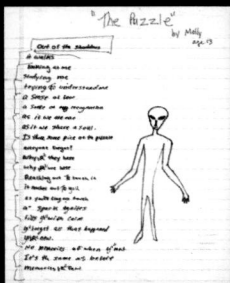

'The Puzzle' by Molly, age 13

Out of the shadows
it walks
looking at me
studying me
trying to understand
a sense of fear
a sense of recognition
as if we are one
as if we share a soul
Is there some piece of the puzzle
everyone forgot?
why are they here
why are we here
Reaching out to touch it
it reaches out to you
as your fingers touch
a spark ignites
fills you with calm
you forgot all that happened
your new.
no memories of when you met
It's the same as before
memories are few

"The Puzzle"

by Molly
age 13

Out of the shaddows
it walks
looking at me
studying me
trying to understand me
a sense of fear
a sense of recognation
as if we are one
as if we share a soul.
Is there some pice of the puzzle
everyone forgot?
Why are they here
why are we here
Reaching out to touch it
it reaches out to you
as your fingers touch
a spark egnites
fills you with calm
you forget all that happened
your new.
No Memories of when you met
It's the same as before
Memories are few

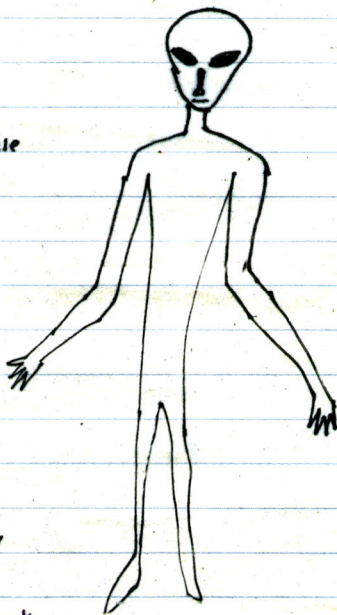

Strange Visitors - East Field in Alton Barnes, Wiltshire, has always been a hot spot for UFO and crop circle activity and the following encounter has only furthered the belief that this area of land is something special.

The two witnesses, Jenny and Steve, were driving by East Field when they noticed a newly formed crop circle. They parked on the farmer's track to have a better look at the formation and noticed six tall figures, wearing hooded cowls, standing inside.

The beings were moving around the circle with their hands above their heads, in a rhythmic manner, from the centre to the outer edge.

The two witnesses decided to stay and wait for the strange visitors to leave the field. It wasn't long before the six figures exited the circle and were heading towards Jenny and Steve's vehicle.

The entities were all over six feet tall, two of the group were male and the other four female. Their faces were long, with large eyes and high cheekbones. Their hair was similar in style to a medieval page-boy, parted in the middle and reaching down to the shoulders. Two of the females were wearing green cowls, a third wore a red cowl and the forth a yellow one.

The witnesses noted that all six of them looked very similar but found the females to have an unnerving, non-expressive manner to their eyes.

One of the females headed towards Jenny's side of the vehicle... the others followed and gathered around. Jenny immediately engaged the entities with questions.

She asked if the circle was fresh and the female answered in the affirmative with an accent described as part Germanic, part Dutch.

Jenny inquired whether they had seen circles before and again the female agreed in the affirmative. Jenny felt the beings weren't of this world so she questioned them about where they were from. She was surprised when the females answered 'Holland'.

Jenny continued with why circles only appear in cereal crops? The females retorted that circles also appeared in vegetables, trees, ice and sand. Jenny probed the group on how they had arrived in Alton Barnes, but that time she received no reply.

Undeterred, Jenny's interrogation persisted... what had they been doing in the field? A range of random answers pursued... "We were testing the circles for vibrations", "We were dowsing with metal rods", "We were feeling the energy under our feet", "We make a study of these circles."

Strange Visitors   Ary by Kim Carlsberg   Andy Russell

Alien Jewel - The female then started digging around inside of her cape in a searching manner and quickly announced "I have something to show you."

Jenny extended her arm in response but was immediately halted. "No, open your palm" the being instructed.

Jenny turned her out-stretched hand to reveal her palm. The stranger's hand darted from her protective sleeve to retrieve something from within her cloak. Even though the movement was very fast Jenny noticed that the fingers were very long, but never saw a thumb.

The Jewel - The visitor quickly placed a peculiar, circular, metallic object directly onto Jenny's palm. The underside of it was rounded and seemed to hug Jenny's skin tightly. The metal was shiny and tapered at one end.

On the upper surface there were hundreds of facet cuts and on top of this was a motif similar to a fern leaf but with a fractal style, like ice frozen on glass. There also appeared to be many tiny holes. Jenny

was transfixed by the mysterious jewel for several minutes before passing it to Steve who commented on how extremely light it was.

Jenny started to hand the article back to the females when again, she was interrupted and given specific instructions to simply hold it flat in her hand. Jenny obeyed, and the female's hand shot out and swiftly retrieve the metal medallion.

The female explained that it had been discovered in a crop formation in Holland and after scientific analysis it was decided that the object was not from Earth.

The females walked away and then suddenly disappeared, leaving the remaining two males. Jenny queried them as to what they thought this all meant.

They replied in turn, in a heavy Germanic accent. "They are trying to tell us something, we will not listen. We must stop pillaging the earth." "We must stop exploiting each other and stop killing each other or God will be very angry."

Then one asked "Did you know the DNA of the wheat that has been flattened is different from the wheat that is standing next to it?" Jenny replied that she did and this answer surprised them. "We must go now." They then proceeded to walk past the vehicle and then simply vanished like the others.

The witnesses decided to explore the crop circles and discovered there were blackened sections at the centre of the circle.

Alien Jewel   Art by Kim Carlsberg   Andy Russell

Touched - "My mind was blank, the silence deafening. I saw nothing but a swirling black hole above my bed and the shimmer of thousands of stars being warped and twisted through this strange and alien portal.

I outstretched my hand, into the darkness beside me, encountering a strange object. Something soft and warm. Fear began to radiate through me as a hand gently took mine, a hand that was not supposed to be there. I strained to turn my face to see who stood in the shadows beside me, but my muscles refused to obey. As the delicate fingers caressed my palm, I felt my fear dissipate and fade away.

A wave of calm and comfort spread over me, the safety one feels as a child being held by their mother. I closed my eyes and felt the all encompassing warmth turn into a vibration as I melted into the night. "

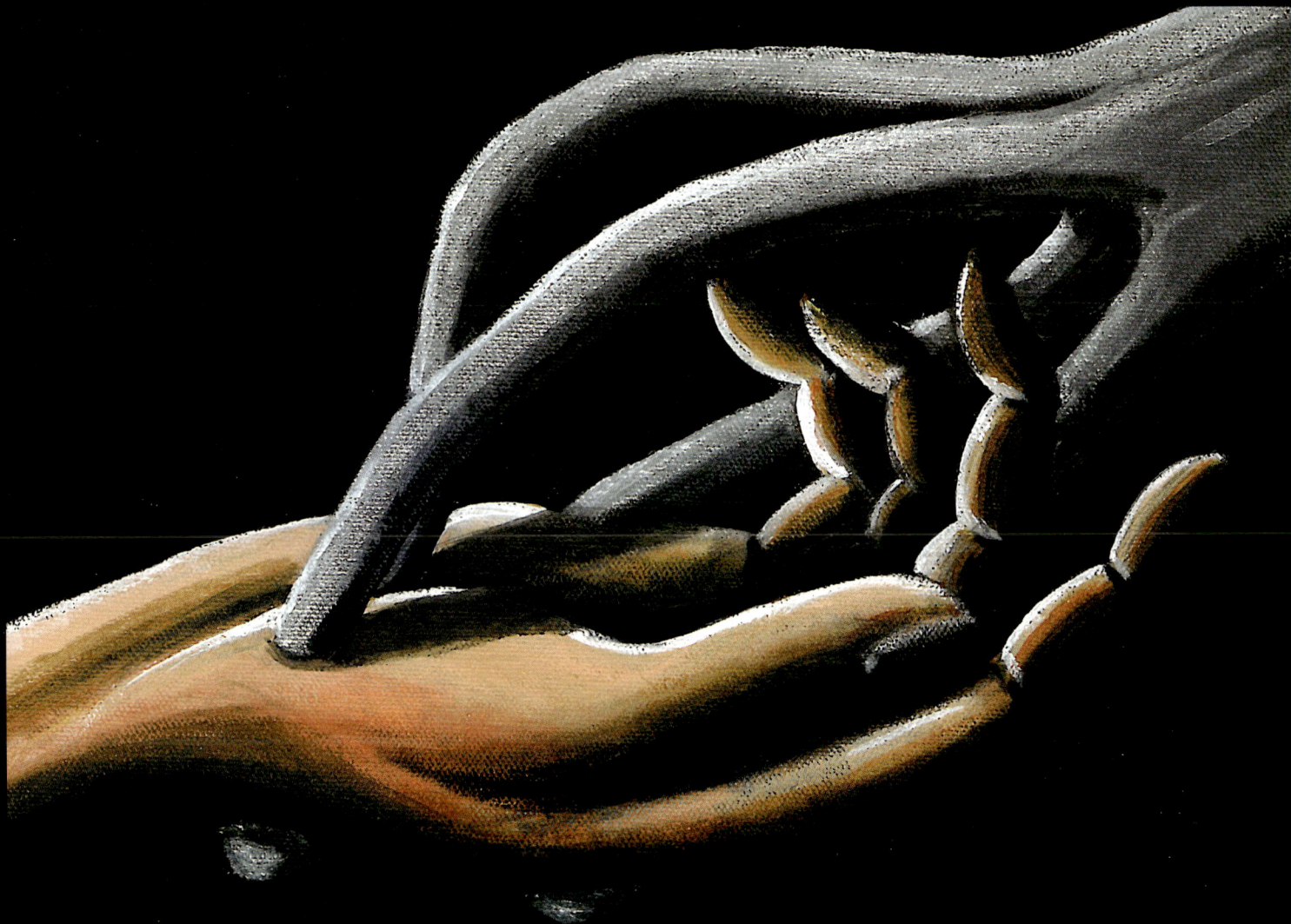

Touched

Barbara Gunderson

The Blue Beings - We are called many things... those of us who claim contact with ETs. Some don't believe us, or the possible of having encounters with beings who can't possibly exist. In extreme cases, skeptics and/or psychologists label us with psychopathologies.

Yet, here we are, human for all intents and purposes (perhaps some hybrids in the mix), and if we are so led to tell other humans about our other-worldly encounters, then what better way than through art? Our numbers are strong, and as we come forward and present the language of the visual, we give others the opportunity look directly into the eyes and energies we have engaged.

The ET encounters and artwork began for me in 1989 when I was twenty nine and very busy with my companies. I had my first "dream" about a large, wedge-shaped spacecraft.

My background was in horses, art, music, dance and advertising. UFOs, ETs, and sci-fi were not on my list of hobbies! After the spaceship dream, which seemed unusually real, the auto-writing, channeling, and auditory messages began. I quickly learned to communicate with the ETs through psychic classes and meditation. My list of paranormal "incidents" began to grow, and by August of 1990, two tall blue beings, identified to me by the names of Z'hara and Sha'mor, were guiding my paintings, physical changes and "ET

education." I was being prepared to move away from my family, friends and dream of playing drums in a rock and roll band (that was/is UFAUX), for complete immersion into the world of UFOs.

The messages from these beings placed strong emphasis on environment and choices humans had to make regarding whether they would continue with violence and hostility towards each other, or work together to save the planet from complete destruction.

The extreme benevolence and love these "tall blues" have for humanity and Earth is literally overwhelming. They will help us, if we ask. They are virtually angelic, and hence the frequency of "wings" - actually vrili - or streams of energy,in many of my paintings. They also are noted to have dolichocephalous, or elongated skulls. Such skeletons have been found exhibiting these unusual skulls, yet their origins remain unknown.

Along with environmental issues, my contacts stressed the importance of science and human health, particularly the area of non-invasive medicine. The beings lead me to a little-known professional optometric association that practices a healing modality of applying colored light through the eyes.

Since those first encounters began, I have produced a myriad of artwork and lectures that are designated to enlighten, educate, and inspire humans to not only consider our known history, but expand our thinking to encompass otherworldly technologies and beings that are only a very minute distance from our current reality.

My paintings convey the future through symbols, color and light and represent a very reachable, beautiful, peaceful time for humanity, if we can only grasp it all before it's too late.

Blue Beings   Susan Gordon

Slanted Eyes - In the mid 1950s, while living in Compton California, I woke up in the middle of the night to see a little, bald, slant-eyed extraterrestrial looking at me through the living room window.

I had been asleep on the sofa, because my visiting cousin was in my bed, so I was all alone without my three sisters.

I was only five years old and began calling out... "Mommy, Daddy! Mommy, Daddy!" No one heard me. I was too frightened to leave the sofa and run back to the bedrooms, so all I could think to do was pull the covers up over my head.

The next thing I knew, in what seemed like only five minutes, I opened my eyes to see the sun coming up. I was still laying on the sofa, but I was on top of the covers and turned around so that my head was where my feet had been.

In the mid 1980s, I decided to see a psychologist who was a trained hypnotherapist. During the regression, we learned that this little ET had a companion, a tall beautiful female ET. She was very human looking with long, wavy hair and a light brown complexion. The two of them had come for me.

I was terrified when the tall ET approached me on the sofa but as she slowly reached out and touched me, instantly, all of my fear disappeared. I felt total love from her and the little slant-eyed ET. I could even hear their thoughts. They said they were very happy to see me, again.

The beings were very loving and took me aboard their ship. I was given a healing that was administered with pretty, pastel colored lights of peach, lavender, and mint green.

The tall, beautiful, female extraterrestrial was very nurturing, like a mother to me. She allowed me to play and run up and down the corridors of the ship. She didn't have to walk, she effortlessly gliding behind me as I ran faster and faster.

When it was time for them to return me to the sofa, they had to remove my memory of the ship because I didn't want to come back to Earth. Can you blame me?

Slanted Eyes   Art by Corey Wolfe   Denyse Baham

Shimmering Lights - An extraordinary event occurred in my life a little over thirty years ago. At the time, I was living aboard a sailboat, a forty-seven-foot trimaran. This modern three-hulled vessel had been built by my partner and me in Portland, Oregon, and sailed down the Oregon-California coast in the spring of 1968.

After several months in San Diego, California, we sailed on to Hawaii, a voyage of over twenty-four hundred miles. This sailboat was our home for over two years. Early one morning just a couple of minutes after 5:00 a.m., I looked out the open skylight from my bunk at the beautiful early morning sky. The sailboat was tied to a dock in the Alawi Yacht Harbor in Honolulu, Hawaii. I had gotten up a littler earlier, dressed, made coffee, and then returned to stretch out on my bunk. My mind was filled with the events of the coming day. My partner and I had just sold the boat, and the new owner would take possession within a few hours.

Suddenly, I experienced a dazzling array of bright shimmering light followed by an intense tingling of energy, which filled the entire stateroom and engulfed me. My entire body began to vibrate as a beautiful woman materialized in the walkway right next to my bed. My first reaction was, of course, to believe that I was dreaming or perhaps hallucinating.

She spoke to me and took my hand. In a matter of moments, I came to understand that she was real and much more alive than I. The energy aura that she emanated made my entire being vibrate at a level that could only be described as ecstasy. I asked, "Why have you come to visit me?" She replied, "Myself and others like me are friends, and we want you to come to a very important briefing." Moments after agreeing to go with her, an unparalleled series of events occurred. First, I was teleported to New York City. I then boarded a saucer-like spacecraft and was transported to an extraterrestrial location for the briefing. The extraterrestrials orchestrated the briefing to give me and about fifty other guests a glimpse of Earth's history. They also showed us a dramatic and vivid view of humankind's potential destiny.

After the briefing, I was asked to undertake a mission. I then agreed to publish a book "Safespace" about my experience and to produce a motion picture feature which would tell the story of my briefing and help the people of planet Earth understand and change their destiny. Upon returning to the sailboat, I realized that only thirty-three minutes had passed. However, from my personal perspective, it seemed like many unhurried hours had been spent with these incredible beings. To this day I remain convinced that the series of events were real and not a dream or delusion.

The Federation Starship "Titha" on the next page is extracted from that story, "Safespace", wherein an extraterrestrial "Regon Varce" is sent to save the people of Earth from destruction. He also reveals why Earth is such a special planet and why its existence and location is such a carefully hidden secret. If "Safespace" is successful, Earth and all its people will survive and discover who they really are, and why the Earth is the key to "the Federation's" own existence.

Shimmering Lights   Robert Miles

Semjase - Edward "Billy" Meier, a one-armed, Swiss farmer with a sixth grade education claimed repeated contact, beginning in 1975, with a lithe, enigmatic female cosmonaut named, Semjase, from the Pleiades star cluster.

The stated intent of her visits was to impart spiritual insights and metaphysical truths—to remind Earth humans that there were other "rational, thinking beings in the universe." Soft spoken and petite, her skin was described as flawless porcelain white, her eyes... pale blue, and her blonde hair flowed shoulder length. She conversed with "Billy" in his native Austrian-German tongue.

Aside from being a skilled star pilot, she was a master of multiple science and sociological disciplines. A youthful 300 years of age, Semjase said Pleiadians live to be more than 1000 years old. Whether real or a figment of Billy's imagination, Semjase's description is characteristic of many reports of beautiful humanoid beings that have come to represent the "Pleiadians."

Semjase

Jim Nichols

PROBES HAPPEN...

Sign Of The Times - My UFO experiences started at an already vulnerable period in my life. It was shortly after I was healed from cancer, my father's death and an unwanted divorce. During my cancer treatment in high school I watched the news, weeped for kids being murdered and was deeply into comic books. I prayed to be a super-hero to help take the pain off the Earth. I had dreams of a shaman visiting me during treatment and I went into remission.

My father passed during that time, and I divorced an Italian woman whose family was mafia. They unsuccessfully tried to kidnap her on our wedding day, then quickly forced us to divorce. It was all too much... I experimented with the occult, partied, and lived life on the edge like the trickster fool.

It was during this time I began to see UFOs. I was messing with the occult learning to cast spells (never to hurt anyone) and reading anything I could get my hands on, trying to figure why I was cursed with so much calamity. Life was hard enough already... throwing in UFOs and weird creatures really sent me over the edge.

Luckily, many of my sightings included witnesses so I knew I wasn't loony. Friends said I must have been attracting these events. I can recall seeing many different types of UFOs; black pyramids coming out of a green-orange vortex with helicopters chasing them, a craft over a vehicle I was in, and an orange, rusty-type sphere floating over an intersection I was walking through.

At one point, someone (a government agency I'm quite sure), monitored me and my home. A blade-less, soundless helicopter appeared with a man in a military uniform in the cockpit who actually waved at me and my companion. I once saw a golden ball of light descend from a craft as if to get a closer look at my friends.

The attention was more than I could handle. I became a hermit, read all I could on the subject, then eventually built my website and co-created a radio show to further the research and raise awareness. If it weren't for the host of books, documentaries and people who have shared their similar experiences, I don't know where I would be today, possibly in an asylum.

And the beings! I have seen beings of light that looked like angels, serpent like entities - reptilians perhaps... unforgettable, strange, terrifying... yet beautiful. I received telepathic messages from them telling me to write and sing and that they loved me. Even to this day some sort of connection still exists.

I've witnesses small beings in black robes standing around a black box in the woods, three beings standing by a telephone pole - one with a dragon head, a beautiful blonde haired lady and a man sitting Indian style playing a flute... levitating! That trio scared the hell out of me then, but now I wish I could see them again and again. I still catch sight of an occasionally UFO, but nothing like the days of old. I believe when I prayed to be a super-hero, God opened my eyes and showed me around a little bit, but I definitely got more than I bargained for.

PROBES HAPPEN...

Sign Of The Times   Art by DW Freeman   Jeffery Pritchett

The Little Monks - In 1994 I taught a general interest night school course on UFOs at a local high school in Toronto, Canada.

One of the things our class did on a week to week basis was review a brief encyclopedia of UFO terms, names & events. When we got to the 'bedroom apparitions' topic it inspired one of my students, a female in her 30s, to draw the image on top of my hand-out.

She didn't give me the drawing until the end of the class and on seeing it I almost fell off the desk I was sitting on. My student claims that this is what appeared to her one night and she has dated it 'mid Aug 87'. These four little 'monk like' beings appeared at the foot of her bed. She doesn't remember if anything else transpired but when they left, they just went through the bedroom wall.

The reason why this drawing is of importance to me is because in the fall of '88 I had a similar 'little monk' or 'Friar Tuck' character appear in my bedroom in the middle of the night. I could sense something in my bedroom and when I opened my eyes I thought I could see a shrouded face at the foot of my bed. My bed did not have a foot board, and the face was just above the blanket. I couldn't make out any facial features. At the point where I was about to ask 'who are you, what do you want?' I thought I saw a pair of hands that were moving in a motion that meant 'calm down, everything will be ok'. With that, the being just evaporated. I wasn't frightened but my heart was racing and the hair on the back of my head was raised, and I quickly broke out into a sweat.

The whole sequence might have lasted 10 to 20 seconds. What is really of more importance in what I saw, is that my mother died two days later. She had been in intensive care for three months, and on the evening before my apparition she mouthed the words to me 'I'm going to die, aren't I'. I didn't have the courage to nod my head yes.

I don't know if what happened to me in '88 is of a ufological or spiritual nature, or both. But I think about this picture often, and wonder what things are all about. What is reality and where are we all going? I do have vague memories of childhood abductions and being involved in a 'secret school'.

After I saw my first UFO in Sept '75, I've been involved in their research ever since. I've seen three more since then. The last one might be 'one of ours'. I grew tired quickly of the UFO groups. I didn't want to collect or measure more data; I wanted to understand what we were really dealing with.

In the last 10 years I've moved into the exopolitical arena, in an effort to help people prepare for the moment of disclosure or 'knowing'. Interestingly my wife and I recently watched the film 'Knowing' with Nicholas Cage. I liked the way the film tries to tie spirituality and ufology together.

UFO Encyclopedia - B

Ball Lightning
Barclay, John M.
Barnett, Grady Landon
Barry, Bob
Bases of UFO's
★ Bedroom Apparitions — ★
Benitez, Juan Jose
Bentwaters/Lakenheath
Bermuda Triangle and UFO's
Bernstein, Prof. Jeremy
Biot, Dr. Maurice
Blackouts(Electric grid failures)
Blue Book, Project
★Boas, Antonio Villas
Bonilla, Jose A. Y.
Breen, John
Broadlands
Brown, Thomas Townsend
Bufora
★Bush, Dr Vannevar
Butler, Alex
BVM Connection

G:\data\wp51\mike\ufo\course\encyc_b.txt

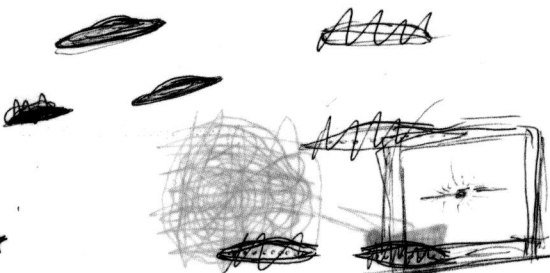

The Little Monks   Mike Bird

Alien And Human - The first alien I remember speaking to appeared as a giant snake sitting on a stump in the woods, that communicated to me telepathically. I must have been eleven at the time.

When I thought about the conversation later I knew that it couldn't have been a real snake. First, because the subject of the conversation was very human-like, a snake wouldn't have known or cared about where I lived or what I did at school. Besides, I remember its large hypnotic eyes only inches from my face.

I had many encounters in the years that followed, but most of the memories are somewhat shrouded or distorted, for my own protection I guess. Occasionally I was allowed tiny brief glimpses, but, because they caused stabs of almost painful fear, the cloaking of some sort was put back in place immediately.

On one such occasion, an alien who was behind me took my hand and told me not to turn around. His hand, however, was enough to send a horrific chill throughout my spine, but for a split second my curiosity took over and I quickly turned around.

I saw a horribly, scary, dark insect head and turned away instantly. He did not look like the typical images of the grays. He did have large black eyes, yes, but the shape of the body, the shoulders, the neck, everything was much more insect like with greater defined ridges at the joints. There also appeared to be some antennae like hoses and possibly mandibles sticking out of the head.

The alien felt bad about scaring me and crouched down in a ball. I felt guilty about making him feel bad. Then he proceeded to shape-shift and in seconds turned into a perfectly acceptable man. He laughed and said, "Is this better?"

I tried to draw this alien later, but unsuccessfully, because I only had the chance to look at him for a fraction of a second.

Aien And Human

Rita Andreeva

Alien - I have a couple more memories of the conventional grays, where they actually looked like they are normally described. These experiences were also cloaked and appeared as if dreams.

One of them was lots of fun: they took me up on a UFO and told me it was one of my lessons, and then they made the floor disappear and told me to get across.

I've always been afraid of heights, so I found some beam under the ceiling, grabbed it and tentatively tried to hang from it. To my surprise I found that it took no strength at all because I was almost weightless, so I laughed and crossed the UFO twice.

The aliens were pleased with me and brought me back down and let me off in a park near my apartment building.

The memories I have that are solid, not at all dream-like or distorted are, unfortunately, only of the lights in the sky and of a blue ray of light.

I think the reason for the dream-like cloaking, is that the human brain cannot handle too much strangeness. As far as the government is concerned,

I have no theories why they still want to keep UFOs a secret.

Probably just individuals who didn't want to appear stupid... although, it seems much more stupid to me to say that God exists, since there is a lot less evidence of that.

And so many government people aren't ashamed to admit to being Christians, if fact, it seems to be a requirement for being elected. LOL. Didn't Obama have to convert to Christianity in order to get into politics?

I can't think of any intelligent reason for the government to hide the information about UFOs and aliens, but then "intelligent government" really is an oxymoron. It's pretty ridiculous trying to keep it a secret anyway... when I wanted to find UFOs I drove into the mountains alone every clear evening for two months straight, and I did see a lot of interesting stuff.

The government can't do anything about the UFO activity eithor. They've tried shooting at them, they've persecuting people who have seen them, whatever, everything to no avail. I put my money on the UFOs for sure. Ha, ha.

Alien   Rita Andreeva

Nordic And Friends - The things she spoke of were deeply intertwined with Hopi culture, yet she asked me what I thought about them. I didn't ask her any questions as I thought that it would have been rude.

The girl, whose name was Debbie, began to speak in Hopi to her sister. Suddenly, from their foreign language, came a familiar word. They had spoken it in Hopi, but to my amazement I recognized it. I immediately knew it as one of the words the Aliens had spoken to me many times in the past.

Debbie continued playing in the afternoon sun, not knowing how much her words had affected me. I asked her what the word meant. She told me it meant "thank-you"—a woman saying thank-you in Hopi. I was stunned, as this word had always remained crystal clear in my mind since they used it to communicate with me in 1988.

After Debbie told me what it meant I realized that the voice I heard speaking in this language at the time I wrote it down was female. My God, I thought, how far will this take me? How much do I need to see, hear and know before I stop questioning everything, and simply believe? After sitting in the blazing sun for a short time I got very thirsty.

I asked Debbie and her sister if they wanted to go to the store with me. They eagerly said yes and we walked over to a small shop on the Mesa to purchase a drink. Once inside I began a conversation with a Hopi man during which I felt a strong urge to turn and look around the room.

"What are you doing here"? I heard in my mind. Standing at the counter with his head turned and looking straight at me was a Nordic Blond Alien! With one glance one could tell that he was not normal looking. He stood about 6' 5" and had broad shoulders. He was very athletically fit and wearing a t-shirt and blue jeans. His brilliant blue eyes seemed to swim around the room. His hair was true white and straight, falling to the middle of his neck, while his pale, almost transparent skin covered his well-defined cheekbones.

"What do you mean what am I doing here?" I snapped back at him through my mind. Again he asked, "Why are you here?' He didn't take his brilliant blue eyes off me as he waited for an answer. I looked at all the Hopi in the room, Debbie, her sister, the Hopi man and the woman behind the counter. None of them paid any attention!

Nordic And Friends   Art by Corey Wolfe   Miriam Delicado

Playing Amongst the Stars - One can still remember all the fun we had! Awaking in the middle of the night, climbing out of the window, the tree-tops blending into a blanket of green as the valley opened up below, and the unforgettable brilliant blue light, was all so surreal.

We were all children, laughing and playing aboard the ships. Hide-and-seek is still the favorite remembered game! Sometimes we had long periods of playtime together. Other times we were separated, to independently learn on our own.

We were introduced to many different-looking star people. They taught us about the Earth, the stars, and the grand cosmic cycles. They instructed us about frequency, vibration, and light. They also shared withus the vast knowledge of star civilizations. They said the universe is teeming with life! But most importantly, they taught us how to love and showed us that everything is an expression of the divine cosmic ntelligence - the Creator of All.

The words of one star being still resonates clearly, "The Earth is a womb planet which is designed to seed life. It is a planet of diversified life-forms where many star races originate before venturing out into the galaxy.

What occurs on Earth will affect the entire universe, as a whole. Thus is the reason so many are interested in the current affairs at this time. We are all inter-connected as such... "The Law of One."

Reaching   Art by Roy Young   Tonia Madenford

Rising - The knowledge they shared seems so complicated for a young child's mind. Yet they took us by the hand and made it all so simple to comprehend. Many times, no words would even be said, just images and thoughts placed into our minds. They were our teachers, but embraced us as their own kin. It was a nursery of learning, nestled deep within space.

This program of global stewardship has been implemented throughout history. It is a mentoring process to assist humanity through its evolutionary stages of development.

It is about raising consciousness. It is about accelerating awareness. It is about imparting a greater understanding of the divine cosmic force. We, as young children, were nurtured, guided and educated in this way.

There are many star civilizations here assisting this planet with its transition into the galactic society. They are here helping and inspiring humanity to advance as a space-faring civilization, graduating as a harmonious member of the cosmic community.

When one looks up at the night sky in total silence, the laughter of children can still be heard. The curiosity of those times still linger... a dashing memory, a fleeting emotion, a warm fuzzy feeling, an inner reflective thought, and the extraordinary experience of wisdom gifted while playing amongst the stars.

Rising   Art by Roy Young   Tonia Madenford

Psychic Encounter Art

Bryan's Journey - began in California and has taken him to live and work in many magical places, such as Sedona, Mount Shasta, Hawaii, Australia and Las Vegas. Bryan was contacted by several groups of extraterrestrials and awakened to his destiny path on December 4, 1996. This awakening coincided with the arrival of the Hale-Bopp comet and the opening of a secret chamber of the Cheops pyramid in Egypt. Utilizing artistic talents from previous incarnations, he has channeled over 12,000 images including various galactic blueprints, sacred geometry configurations and glimpses of other worlds and dimensions.

These channeled art pieces or accelerators are created within minutes and are energetically infused with many levels of information. All images are made manifest using an interdimensional acceleration drawing technique, which Bryan has perfected in previous incarnations. The originals are all hand drawn on 19 x 25 paper using Prismacolor pencils and metallic markers.

These images or accelerators contain ancient and interplanetary scripts and symbols, which are personal and universal keys to forgotten memories and advanced knowledge. Bryan's contacts continue to occur on a regular basis and are both telepathic and physical, depending on which group is connecting with him. The goal of Bryan's artwork is to refine and master the art of ascension as well as to assist others in assimilating and grounding the 5th dimensional energies of the coming Golden Age.

Through The Looking Glass - Contact Date: March 1997 - Drawing Time: 15 minutes. I awakened early one morning in 1997 and sensed a strange electrical charge in the air around my home. As with my first few contacts, I knew what was happening and quickly sat down to render the detailed vision coming forth in my mind. As I began drawing it seemed as though there were, all of a sudden, multiple images linking themselves together in my mind like a giant puzzle.

About a week later, to my surprise, I realized this was to be a huge, 72-piece blueprint which represents the visual story of this planet's history and its passage through the 2012 time-gate. In its totality, this art-piece is truly like a giant, wondrous, multidimensional peak at our planet's history through the eyes of our extraterrestrial ancestors.

This particular image, Through the Looking Glass, is linked to our original planetary creators and the Inter-dimensional Association of Worlds. It depicts the hand of one of the creator gods, our ancestors, magically forming the energy of our planetary sphere. It is also a representation of the journey of many life-forms and worlds, passing through vast time-gates and wormholes to new destinations throughout the universe. Looking at this image awakens cellular memory and initiates the extended DNA sequences within our bodies. Additionally, it connects us with the master races and programmers who originally seeded our planet with a myriad of life forms from other galaxies and ethereal worlds.

Through The Looking Glass   Bryan DeFlores

Blue Starfire - Contact Date: August 2006. Once again, I was awakened around 3:00 am and felt the presence of the extraterrestrial visitors in my home.

As I walked through the house my mind began receiving telepathic signals as well as a visual transmission depicting three tiny star beings swimming in a beautiful sea of blue energy ribbons and cosmic stars.

As I sat down to draw this vision, my whole body was surrounded with this energy, which I could both see with my physical eyes, and feel upon my skin. Although the image was completed in a matter of minutes, this amazing energy stayed in my home for more than a few weeks.

Further telepathic messages and a detailed description arrived after the image was completed. Also known as 'The Blue Essence' and as

'The Flame of Creation', this legendary and very elusive substance appears during all initial creational moments. This includes the births of all beings, planets and universes.

The Blue Star Fire essence still remains a mystery to even the most advanced races in our universe and usually is only detected through one's inner vision. This image depicts and is infused with 100% of the Blue Star Fire essence.

Contemplating this image will allow you to access and align with the magical gifts and talents you have known in former incarnations.

In addition, this Blue Starfire image can be used during the initial stages of the creation of any project or endeavor as it will enhance creativity, focus and passion, as well as add in a touch of magic and mystery.

Blue Starfire   Bryan DeFlores

The 'Money Tree' Of Life - Contact Date: November 2007. I always look upon this image with a sense of wonderment and adventure, as it came through at a point in my life when I was being taken on many astral journeys by my extraterrestrial friends.

I actually have no recall of drawing this image as it was created in a higher dimension of this planet, which is like a fantastic dream-world where all things are possible.

Up to this point in my contacts I had been extensively trained in spiritual remote-viewing as well as teleportation and time-travel processes, however this journey required only a child-like sense of joy and innocence.

The description of this image is as follows. Within the higher dimensions of Earth, there lie many wonders accessible only through our spiritual perception and inner 'mind's eye'. One of these wondrous creations is known as the

'Tree of Unlimited Prosperity'. Throughout the eons of time, individuals have been accessing this divine living creation and anchoring its seed within their mind and heart, allowing its beauty and abundance to flow into their lives in an ever-constant stream.

Truly, this 'prosperity tree' is accessible to all who wish to bring the perpetual flow of unlimited financial abundance into their lives.

To call forth this 'Money' Tree of Life, all you need to do is connect with this image two or three times a day. Ask that a seed of its divine energy begin to grow within your life and create an ever-flowing 'money tree' just for you.

Secondly, be aware that this divine tree will prompt you to do certain things within your life to help it grow and open new avenues for the money to flow through. In addition, the concept of giving the gift of money to others will further extend your prosperity as you go through life.

Money Tree

Bryan DeFlores

251

Arcturan Matter Transmitter - My name is Elliott Maynard. I am a Scientist, Author, and Global Ecologist, currently residing in Sedona, Arizona. I have created over one hundred futuristic artworks, which include paintings, linoleum block prints, woodcuts, murals, wood and welded metal sculptures.

My unique style of artistic expression blends a high level of psychic intuition with leading-edged technologies in fresh new ways. I created the term, "Future Science Art" to describe this innovative new process, and refers to my creations as "Interdimensional Living Artforms."

Although the majority of my artworks would seem to fall into the category of futuristic fantasy art, other pieces depict whimsical animals and bizarre sea creatures. I have also created abstract sculptural assemblages which I call "Imaginariums."

Many years ago, I requested some help from my spiritual guides, so that certain new energies and concepts could be brought down through me into the physical realm, in the form of unique artworks.

It is my firm conviction that some of the most promising solutions to our present "Global Dilemma" lies in humans becoming open to energies and inspiration from our extraterrestrial allies.

Recently, I had a session with a long-time associate, Terra Sonora, who is a professional trance-channeler for several different groups of light beings in the higher realms.

This is what the Arcturian Council had to say about my new paradigm of Future Science Art: "With regard to your Future Science Art, this is a technology you have channeled that awakens Extraterrestrial Awareness in the consciousness of others. Your artworks indeed carry the frequencies of the stars, and helps awaken the Star Beings that are asleep in human form at this time. That is one of the intentions of the artworks you have created. It is to awaken the higher consciousness in the form of "Stellar Awareness."

This sculpture came about when I discovered the metal cover in a local transmission shop in Sedona. I used to stop by occasionally with a six-pack of beer in exchange for being able to root through their junk piles of gears and transmission parts.

The center piece was simple to add with the rainbow art-glass sphere, and the moss-covered red rock from my property, provided a base. The title simply came to me via my intuitive senses, and thus this strange creation was born.

Arcturan Matter Transmitter

Elliott Maynard

These universal art children, with their special quality of "livingness," can reach out to other individuals who need love, companionship, and special beauty in their lives. By the very nature of their creation they have the power to radiate their energetic signatures, and teach us - through what might best be termed, "direct consciousness transfer."

My overall objective for this project was to create "a critical mass" of over 100 Future Science artworks. With the completion of this objective, I feel that a new and evolved type of creative consciousness has now been established for artists everywhere, now, and for the generations yet to come.

Lyran Dream Generator - Once I put myself into a contemplative state, and started playing around, arranging the junk metal scraps I had on hand, this was one of those sculptures that just simply seemed to assemble itself.

I made several visits to a local welder to root through his junk pile and dig out selected pieces. He only charged me for the weight of the metal. I used a metal bender to form the bowed sides of the arch from steel bar stock, then added some cut nails and art glass marbles.

Piece by piece this sculpture took shape, and the end result was something definitely unique and alien. I have been told by at least two different psychics that this sculpture operates interdimentionally, with pulsing lights and fields.

I developed Future Science Art as a transformative new paradigm for creative artistic expression.

I believe that, with the right attitude and focus, anyone can learn to create and experience art as a healing, enlightening, and self-directed evolutionary process.

In this sense, Future-Science Artworks can be considered as "catalytic agents" for enlightenment, healing, and ultimately, for the evolution of human consciousness. This unique process transforms the artist, as well as those who view and bond with these artworks.

Future-Science Art is a paradigm for co-evolution and transformation, and can thus serve as a truly wondrous vehicle for experiencing and participating in human and planetary evolution.

As a conscious co-creator, the artist works in harmony with the subtle energies of the universe. Within this new and exciting paradigm, lies a very special gift, as anyone with the proper mindset can create their own unique family of "art children."

Lyran Dream Generator   Elliott Maynard

The "IS" - This will have to remain a mystery. It is painted for a particular race of ETs who are living as humans, but without conscious memory of that fact. There are quite a few aliens who have agreed to have their memories wiped when they walk-in, or incarnate as a human being on Earth.

The "IS" is designed to help awaken something inside of you, if you are indeed, one of them. Once this is awake, don't be surprised if you meet a very interesting person and get a very interesting offer.

Hints: There is only one smiling sphinx. The AA was established by this same group of ETs.

The Sphinx is smiling.
The Ships are moving in a caravan.
The Portals are open.

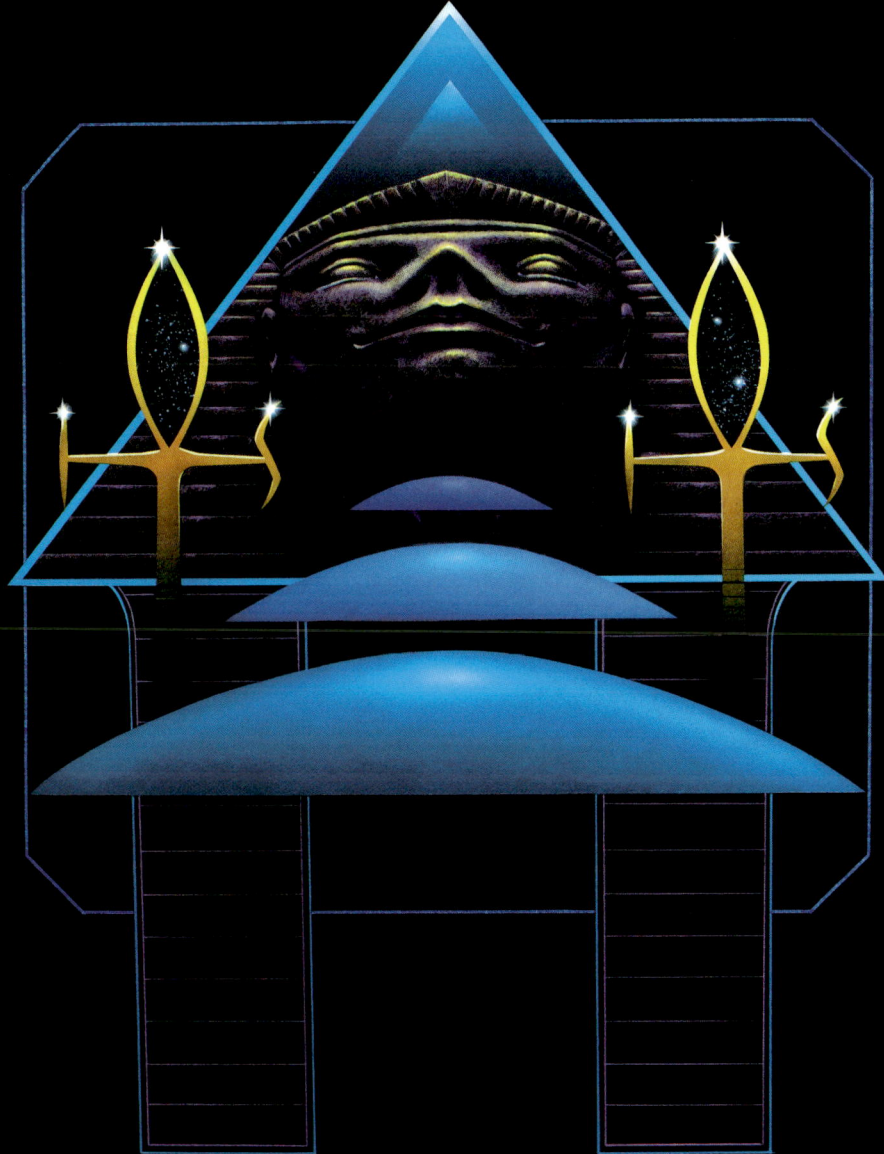

The OS

Corey Wolfe

Inside A Very Big Ship - Back in 1991 I had one
of those "this is more real than reality" experiences.

I was wide awake in bed, yet I was also walking with
a few people into what felt a bit like an airport
terminal. A tall good-looking male (a Nordic)
welcomed me with a sweep of his upturned hand.
I was so excited and knew this wouldn't last long.

I rushed to the view window and somehow
understood that I was actually inside a ship so large
that mother ships were scattered here and there.

The illustration shows a doughnut shaped one
'powering up'. I didn't stay long at the window;
I was so excited to get deeper inside.

I ran down a hallway that ran off to the right. There
were shops, similar to those in a Vegas casino. I
barely rounded the first corner and I was swept back
to this reality. Darn!

Inside A Very Big Ship

Corey Wolfe

Seven - On 7-7-98, I saw this image in my mind's eye. It seemed very important. What I wrote in my journal follows:

This shape (or the same in reverse) is the shape all matter, frequencies, tones, i.e. everything has to pass through to be a part of this reality, this matrix. Many years later, I was listening to the Coast to Coast am radio show. A researcher from northern Washington state mentioned the same exact symbol. I wrote her name down, but misplaced it.

If anyone has seen this, please let me know anything you have learned. I don't know why, but this symbol still haunts me after all these years.

Corey Wolfe

Exploring The Coastline - I call my work "cosmic art" because "the cosmos" extends beyond the physical universe, and even beyond the mind's imagination.

The term "cosmos" often has a spiritual connotation perhaps referring to angels or beings that transcend our three dimensional territory. It refers to a place that not only exists just as surely as our material domain, but includes it.

Sometimes experiences in this cosmic realm seem like dreams, but this kingdom is actually more valid than our waking reality. It is a region of existence that transcends time and space. It is not empty or void, but a place rich with beings and happenings. It is sometimes frightening, sometimes peaceful and beautiful, but it is real.

My cosmic art has content that seems eternal in this

already. When I was quite young I had several adventures in this quantum field that were vividly real, where I encountered non-human beings.

It was clear to me they came from other worlds or dimensions. They were intelligent and communicated with me. Sometimes they would fly in from above, and I could glimpse their saucer-like spaceships.

This was before I was old enough to know about aliens and UFOs, so these episodes were not the result of a sci-fi novel or movie.

The non-human's realm of existence did not have the same density of our world which, once again, reinforcing the qualities of a dream.

I have always wondered if I would ever meet any of those beings here on Earth, on solid ground.

Exploring The Coastline   Yuichi Tanabe

Moai Small Planet - My childhood encounters sparked an intense interest in religion as well as science for me. Both are equally important and seek to advance our knowledge.

In some religions there are angels, in some kinds of science there are aliens. Perhaps they are the same, or perhaps not, and perhaps both are someone's interpretation of the beings from my childhood dreams.

In my art, I aspire to bring together aspects of both science and religion; knowledge of the tangible, with faith in the unseen. I strive to present these ideas in a way that is intriguing and beautiful. Some of my pieces are based on mythologies, especially those that suggest that human beings originated on a planet other than Earth.

My interest in science inspires me to include robots, and/or androids, and spaceship technologies. My fascination with the art of ancient cultures has motivated the inclusion of such monuments as the

Moai of Easter Island and Stonehenge, but existing in different realms. Stonehenge was the stimulus for my "Cosmic Dance" series: One day a picture of Stonehenge caught my attention and proceeded to absorb my consciousness.

Within moments the Stonehenge landscape and my consciousness overlapped. A continuous series of images appeared and dissolved repeatedly.

Some scenes were of an enormous, endless oceans, others were clear-air, high mountain panoramas, while still others held the vistas of large cities with towering skyscrapers.

The insistent images each had personalities of their own. I delighted in their animated energy... they bounced, flew, and danced in the heart of my mind's eye.

These pieces of art from the "Cosmic Dance" series are representations of that transcendental experience watching different realms meeting and merging.

Moai Small Planet

Yuichi Tanabe

265

Children of The Phoenix Lights - I would like to share an excerpt from a letter I channeled from Aliebod, a seven year old human/hybrid child on May 26, 1997.

"There are many from here who want contact with their mothers. I would like you to work with us, Elle. Because there are promises made when they are visiting in dreams, there are times when these promises are not kept. I think, because it is only seen by the mothers as a dream, and not real, they pay no attention when they wake up. Perhaps the realization of someone like you speaking for us the hybrid children, to our mothers Elle, can wake them during non-dreams. We want to know what the word, "love" means, coming from the ones called our mothers."

At six years old, I was conscious of my family sleeping down the hall while I would kneel at the upstairs window overlooking the backyard. A beautiful angel with long blonde hair and a flowing white dress would ride down on a moon beam and talk with me. She didn't have wings, but I **knew** she was an angel. These talks went on for almost two years.

Years later, I remembered I was out-of-body at that time. The angel was actually a human/hybrid child named Ariel, and the moon beam was a ray of light from a spacecraft. I went on with my life and forgot all about this first contact until 1977. A guide, named Zack came through on a Ouija Board and I became a conscious channel. Throughout the years, using telempathy and automatic writing, I spent many glorious hours talking with Zack and other guides introduced by him.

I met Bashar, the ET human/hybrid who speaks through Darryl Anka in 1983. He became a mentor and a friend who has shared his first-hand knowledge of the hybrid children. As we talked throughout the years, I began to "re-member" some of the children as my soul connections.

On March 13, 1997, a V-shaped ship, along with several other types of crafts, is sighted over Phoenix, Arizona and seen by thousands of people. The event was globally broadcasted by major news stations and the ships were quickly dubbed, **The Phoenix Lights**. In a conversation with Bashar the next day, he shared that I have been working with the hybrid children since childhood. Ariel has confirmed that the children were on the ship that night.

I channeled the first letter from Allieo on May 8, 1997 for her mother Stephanie. The next letter, channeled two weeks later, was from Aliebod to his mother with the initials GHO.

In Bashar's ancient language, the children are called, the Shalinaya, which means the first to land. However, in their own language, they call themselves the Ya'yl. They live aboard the ship, which now resides above Sedona, Arizona.

I energetically interact with the children daily. Ariel, the spokesperson for the group is here at my side as I write... The children love to go shopping with me. I always know when Julius is here because he slowly starts pushing the shopping cart when I stop moving it. We have so much fun together and I am honored to call them my galactic family.

Children of the Phoenix Lights   Art by Kim Carlsberg   Elle Keith

Onspired Encounter Art

Saucer Surfing - I usually do more serious UFO scenes, because I take the subject of the presence of the many extraterrestrial races that are visiting and/or cohabitating on our planet quite seriously... but this image is just about having fun.

Since I was a kid I wished I could fly and I have had flying dreams all my life. In this photo manipu- lation I combined a photograph of clouds, an image of the "Sports Model" flying saucer - made famous by Bob Lazar, the Nuclear Physicist who was hired to back-engineer this type of craft at Groom Lake, aka "Area 51", and a photograph of my daughter, to make the scene.

Maybe kids in the future will have something like this instead of skateboards. Oh what fun until you fall!

Unsuspecting - This image was inspired by the thought of aliens watching us at war and wondering what they must think about our violent natures.

The term "foo fighter" was used by Allied aircraft pilots in World War II to describe various UFOs or mysterious aerial phenomena seen in the skies over both European and Pacific Theater of Operations.

There have been numerous reports of UFOs not only sighted, but actually shutting down US military installations where nuclear weapons were being tested. Latest reports by US Marines in Iraq suggests UFOs are monitoring activities there as well.

Triangular craft have been reported over Bagdad. Do the UFO occupants understand our motives or our disagreements? Would they ever openly intercede? Have they in the past or will they in the future?

Maybe to them, we are nothing more than a spherical board game!

Unsuspecting   Brent Berry

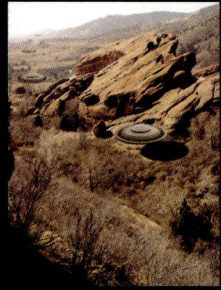

The Gathering - A few miles west of Denver, Colorado, is a beautiful area of geological interest known as Red Rocks Park.

The incredible prehistoric, red sandstone formations there, were once part of an ocean floor about 250 million years ago. Dinosaur tracks from the Jurassic period were found in the area, as well as other prehistoric evidence, including a partial, forty foot sea serpent and fossils of flying reptiles.

Whenever I visit this place I feel like I've gone back in time. The evidence of Earth's greatness and age surrounds you. It's a unique spot and is not very large, but is so physically unique and odd, that it seems like another world. I sometimes imagine that this strange bit of land may look like home to a traveler from another world.

Maybe UFOs have been visiting Earth for millions of years and lay claim to some of these unusual vortexes. I don't know, but regardless, I thought Red Rocks would be a great backdrop for a flying saucer scene, and I was not disappointed.

The Gathering

Brent Berry

Beaming Horseplay - Cow mutilations are common and well known throughout the UFO community.

"Snippy" the horse, was the first horse mutilation I ever heard of. It happened in Colorado in 1967. It was on Harry King's ranch that "Snippy", a three year old, Appaloosa mare, was mutilated. This was the first mutilation to be made public... and I've often wondered about Snippy.

This horse, in my image, has been selected for study and/or experimentation, by an alien craft in the middle of a cold and snowy, winter's night.

Beaming Horseplay   Brent Berry

Rush Hour - Most of us, in this world, have to work
to get by and we have to get there on time. Earthlings
must look pretty stupid to visitors from other places!
Millions of individual machines all moving at once and
so many in the same directions.

Do aliens have a schedule to keep, or does time
matter to them? Are they working too, or just playing?
Maybe aliens stop by on their way home from work
just to get a laugh watching us in our out-dated, oil
consuming, Earthbound, metal mules.

They must wonder if we are ever going to figure out
free-energy, anti-gravidic transportation from all
the old clunkers they have left us in our deserts...
perhaps their covert solution to the universal law of
non-interference.

Rush Hour   Brent Berry

No Where To Hide - Something has gone very wrong at Area 51! Seriously though, what mysteries are studied at Groom Lake/Area 51... ETs, UFOs, Time Travel? Few really know, and only what they need to know.

Imagine with me, if you will... the errors that can occur when humans actually get their hands on these powerful alien technologies. Nuclear Physicist Robert Lazar - hired to back engineer recovered UFOs at Area 51, has stated that he was a "replacement" for one of three men who died while attempting the same feat!

This image plays with the idea of Area 51 employees mishandling a time control device which elicits the aliens to come and investigate the incident.

No Where To Hide   Brent Berry

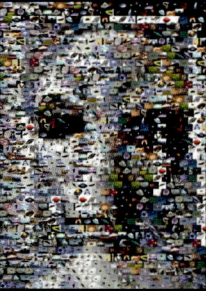

Alien Face - Ever since I was young I have been fascinated by UFOs and aliens. I don't know why. None of my family or friends were ever really into it.

I was never a Sci-Fi movie fan but I did particularly enjoy movies that dealt more with humans and aliens such as Close Encounters of the Third Kind and Signs. When I started creating my montage mosaic art some of my first pieces were alien/UFO related such as the Alien Face.

Though I've done close to 1000 now the Alien Face is still one of my favorites.

Alien Face

Paul Van Scott

Area 51 - I've always been fascinated with UFO and related photos. What are they? Where are they from? Is the US government really hiding something?

The secret government base in the Nevada (US) desert is known as AREA 51. What secrets does this base really hold... I've often wondered. That led me to my extremely popular Area 51 montage mosaic.

When you look at it, are you seeing Area 51 hidden by UFOs or are you seeing UFOs hidden by Area 51?

Area 51   Paul Van Scott

Odd Aunt Shirley - Aunt Shirley was washing her '49 Ford out in front of her house in North Denver when, in the early 50s, my father captured this photograph.

I had a bittersweet relationship with her. The sweet; she was the artist who got me interested in oil painting at a young age and encouraged me to express myself creatively. The bitter; she was an eccentric spinster who, upon discovering that I was gay, disowned me.

I painted the original image in 1972, and sold it to a dear friend, Barbara. She had it hanging in her living room when, a few years later, someone broke into her house and stole it. I was talking to her in 2008 and she still was grieving the loss of her Aunt Shirley

painting, so when I hung up, I grabbed a canvas and went to work. She now has a beautiful Giclee reproduction that adorns her home, gratis.

I attribute a great deal of Aunt Shirley's extremely odd behavior during her lifetime to the possibility that she was an alien abductee.

As the aliens abduct humans intergenerationally... and my father, her brother, was my link to the abduction phenomenon, it is reasonable to assume she was an abductee as well.

It is said that your entire life flashes before your eyes when you die. I often wonder if she saw the little gray with big eyes who's sitting in the front seat of her car when that final moment arrived?

Odd Aunt Shirley

Chuck Chroma

El Receptor de Huevos - or "The Egg Catcher."

I like the title in Spanish because of the duality in
his sombrero/UFO craft in the sky.

This is another of the automatic style drawings
I've been doing lately. I start with a blank canvas
(paper) and draw in the lines that I see because of
shadows and highlights in the blank paper. Then as
I get more lines and shapes an image pops out at me.

These are completely unplanned automatic drawings...
it is a little strange to me because of the imagery.

El Receptor de Huevos   Kristi Kennington

Tres Grises -

The party for them

they say is the day

for them, they

have been taken in

by the weaknesses of our skin;

skinless and naked

joyful and resistant

I refuse to be

one of their contestants.

Saucer Down - The blue crashed saucer image was inspired by the Roswell crash... and the dozens of other reported crashes around the Earth.

Space flight must be very dangerous on the whole... and smallish moons with very eccentric orbits (like Saturn's moon 'Bestla') might pose a more significant threat to passing space ships. I suspect that our solar system, with the hugely attractive aspect of 'life' on our planet, might very well be littered with a vast number of lost saucer wreckages scattered here and there.

Since Jupiter and Saturn have the largest amount of moons and satellites... voila!

Saucer Down

Garth Perfidian

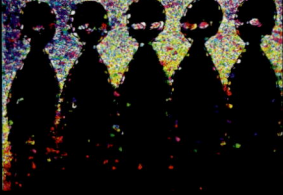

Synchronicity At Roswell - We were driving through Roswell with our land agent friend Glen, with whom we visited the "UFO crash site" and the Roswell UFO Museum.

Glen recounted a close UFO encounter one of our local firemen had some twenty five years ago in Canada.

Apparently the fireman reported his direct encounter with alien beings, for which the fireman was harshly criticized and ostracized. Others were so critical towards the fireman that he eventually became overwhelmed with grief, and unfortu- nately, took his own life.

Glen was most surprised when he saw references to this incident at the Roswell UFO Museum!

Monet Grays At Roswell - This image was created by photographing and digitally manipu- lating hand-drawn renderings of alien encounters on display at the Roswell UFO Museum.

Alien Eye At Roswell - This image was created by photographing the inside of the eye of an alien painting in a store window display at Roswell.

Alien Eye At Roswell

Oilboy

Country Road - This illustration takes me back to my Ohio roots. Despite the Fifties' Hollywood cliché, UFO sightings are not confined to the desert.

I have a cousin in Indiana who woke one night to see a glowing disc silently flying across the field behind her house. The data indicates she is in a growing populous.

This picture is a generic speculation of how a nocturnal UFO might appear gliding down a lonely country road somewhere in the American Midwest.

Country Road

Jim Nichols

The Prank - This image debunks the idea that crop circles were created by drunken mathematicians, by showing aliens could have made them that way too.

The Surf Is Out There - I live and work in Hampshire in the UK which as an artist is most inspiring, not just because of the rural countryside and stunning views, but because of the strange phenomenon... crop circles and UFO sightings.

My work has brought me into contact with Reg Presley (rock star and UFO enthusiast) and I have been lucky enough to meet Colin Andrews – one of the world's leading researchers on crop circles.

As a paragliding pilot I have flown over crop circles and the intricate detailing of these amazing images quite honestly blew me away. Surely, this cannot be made from a human hand???

I am also interested in astronomy and many a night when I am gazing into the midnight sky in the middle of nowhere, I have felt something strange and on more than one occasion have seen something strange as well.

Admittedly, from my imagination and perhaps wishful thinking, my images of extraterrestrials are sometimes media driven and comical or fantastical, but at the core is my own belief that we are not alone in this universe.

Regarding "The Surf Is Out There", as a keen surfer and UFO enthusiast, I wanted to incorporate my two favorite subjects together.

I really hope that if we are lucky enough to receive visitors in our world, they would like to experience some of the fun aspects that our planet has to offer!

... the truth is out there... but so is the surf!

The Surf Is Out There   David Penfound

Extraordinary Encounter Art

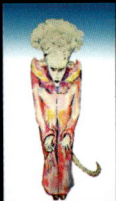

Drakonian - This is an illustration of a Drakonian I encountered who lives below Glastonbury, England – or such as he appeared to me after I explored magic mushrooms as a way to get some questions answered.

As I started to feel the effects, my astral body was whisked away beneath St. John's Square, where I lived at the time. I immediately found myself suspended face-down looking at this incredible creature. I estimate he must have been about fifteen feet tall.

Although chiefly taken by the Anunnaki, my younger daughter and I have encountered many types of beings, but I believe I am meant to help familiarize people with these particular ETs – or in the Drakonian's case - Terrestrial Extras.

Drakonians are hybrids of the Drakons, and human, and were introduced as overseers of the planet, therefore are not ETs, but Earth-bound hybrids.

I use the term TE to differentiate, although strictly speaking, since humanity originated in the dimension above Earth, (hence our 'Fall') – so are we!

We are about to return to this chakra layer over Ascension, the first opportunity we've had to do so

in thousands of years. Some ETs are now assisting us in awaking to this reality a little beforehand.

The Drakonian told me we were ninety meters below ground in what I know to be one of the empty magma chambers of a long-dead volcano.

Glastonbury Tor is the most famous landmark of the town: an artificial mound which covers the central, collapsed stack of the volcano and is the reason Glastonbury (née Avalon) was built where it is. Underground is a large lake/caldera fed by springs, approximately the size of the base of the Tor.

The being poured many images of his extremely long life (8000 years) into my head. He spoke in a deep voice with unusual inflections, it may only have been telepathic, but it was very penetrating. He told me his quarters extended in a boomerang shape between Weary-All Hill and the Tor.

At no point was I able to see the top of his head because so much energy was coming from it. I knew he was able to interact with humans and others of his family over a long distance via the horns on his head he uses as antennae.

After many years of trauma, and still some pretty difficult moments, I now consider myself privileged to have these encounters. I take it as an extremely good sign that an Anunnaki abductee was allowed a rare glimpse of another species that wouldn't usually be perceived - as the Drakon and Anunnaki aren't normally co-operative with each other!

I hope more will be revealed about the various species interacting with us soon.

Fairy Presence - is always strong in sacred places. The presences of place can include gods and angels as well as fairies. You never know what will come through when you invite the invisible residents to make themselves known. I had a lovely and rather unexpected visitation from a fairy while at the Roman ruins at Lydney, Gloucestershire, England. Once this sacred place attracted thousands who came in hopes of meeting the god Nodens, who might heal them as they slept. Many were healed. Lydney has several sites of Roman ruins, including one of the most important camps and temples at Lydney Park believed to be the cult shrine of Sabrina, goddess of the Severn.

The ruins of the dream temple are on private land in a remote location, still wild, quiet and pristine. The Romans built the Lydney healing temple in the late fourth century CE. Their temples were inspired by the dream healing temples of the Greeks, who believed dreams were a way to contact the gods. The Lydney temple was on a much smaller scale than Greek dream temples, but it offered a complete healing sanctuary: a temple for the priests; a hydrotherapy centre of Roman-style cold, warm and hot baths; a dream dormitory or abaton; and stalls for resident physicians, consultants and healers.

Lydney was dedicated to a little-known deity, Nodens. The Romans were great adapters, and Nodens was derived from a Celtic deity: the Welsh Nudd of the Silver Hand (and Lludd, the namesake of Ludgate in London), and the Irish Nuadaof the Silver Hand.

Nodens/Nudd was a water deity, and it is thought the Severn River itself was his silver arm and hand.

To get to the temple ruins, one crosses a small stream (a symbol of a portal crossing into another realm or world) and then climbs a steep, winding hill. As I started my journey I opened my consciousness to drift back two thousand years and sense the sacred temple in its glory days. I was fortunate to have the site all to myself and my travelling companion without spectators disturbing the fragile energy.

After experiencing the ruins, I was attracted to a huge oak tree whose gnarled roots poked above the surface of the ground. The tree itself seemed to be a portal between worlds. I sat down against it and entered into a serene meditative state with my eyes open. I expanded my field of consciousness into the environment with no expectations. Would I meet Nodens, encounter fairies, or simply have a pleasant meditation? I didn't know, but I was open.

After a while, I became aware that I was being regarded by a small figure who had materialized by the tree. It seemed that this masculine presence, came out of the roots of the tree. His clothing was not distinct, but I could clearly make out a vivid red cap on his head. He seemed old, like a little old man, and I had the impression of gray whiskers or a beard. I guessed his height at about two feet. He seemed curious about me, and I knew he wanted me to acknowledge him, for if he had wished to watch me in secrecy he could have easily done so. I gave him a mental greeting thanking him for joining me. I received a mental impression of a greeting in return, along with appreciation of my respect for the place. He stayed for a bit, then suddenly he was gone. I had the impression that he disappeared back down the tree roots.

Crop Circles - The argument of whether crop circles are a genuine phenomena, is finally at an end.

With the increasing complexity of designs - frequently featuring crop weaved together - and the speed with which they can appear (often literally in seconds), we can surmise that 'something' other than humans are behind these enigmatic creations.

More and more we are seeing UFOs, strange orbs and fleeting non-human figures near the circles, either prior to - or just after - the circles appear. With the weight of the current evidence, I believe we can look to an off-world answer to who these mysterious circle creators might be.

Not only are their creations truly fascinating, but the messages encoded in these designs are awe - inspiring as well. Recently, a crop circle predicted a solar eruption, pin-pointing a specific day for this to occur.

The prediction was correct, but what made this even more intriguing is, that the sun is currently at its quietest in almost a century and no solar activity was expected. Surely, only an off-world astro-physicist could predict an event with such a degree of accuracy!

The circles are not only giving us messages, I have heard of so many people coming into crop circles receiving spontaneous healings, deep spiritual insights and literally life changing experiences.

The circles are a catalyst for us all to experience, it would be foolish for us to underestimate these messages and their creators, as they have much to show us.

Art by Kim Carlsberg

Andy Russell

Winged Creature - In 1999, while living with my mother in the Coral Gables section of Dade County, Florida, I saw a curious entity one day while I was leaving her complex in the early afternoon.

As I pulled out of the parking lot, my attention was immediately drawn to a strange "character" standing in the middle of the road and looking around as if lost.

To this day, I cannot say if the person was male or female, but it was striking in appearance. The figure was very tall and broad, blond haired, and seemed to have a pair of folded wings on its back! It wore sandals and a robe-like tunic.

It seemed powerful, and totally out of place. When it looked around and spotted me in my car, I felt some fear and apprehension.

I looked away, and drove in the opposite direction. When I looked back, the figure had vanished.

Behind The Veil - I remember meditating each night trying to will myself to the other side because I had never been there, and got nowhere, but that would change.

One day I met someone who taught me about shamanic plants. They said plants have been used for centuries as tools to unseat the soul. So one day I had to know, I ingested the plant. On this day the veil was pulled back. Now, where I was once blind, I could see, where I was lost, I knew the way. This experience changed me forever. I will share, that the effects of the plant were immediate. I was outside in the courtyard of my home and I barely had enough time to walk into the bedroom to lay down when I fell completely out of my body.

Many times in meditation I had attempted to will myself to float above my body, but with no success. This was so unexpected, suddenly I was hovering above myself. I had the unusual feeling of being in two places at once. I felt so free and at peace. I took note of everything in the room, the furniture, windows, pictures. The cat noticed me, it perceived my hovering presence. But just as I was getting used to this new scene and new feelings, I was sucked through a tunnel. I was on a fast moving train and I couldn't get off.

I was deposited on the other side of a threshold where I felt connected to everything and everything was open to all, there was no hiding. I felt I knew everything. I needed only to imagine the question and I instantly knew the answer. I could see the physical plane from where I just left. I saw myself still lying there. I was able to perceive everything on a molecular level, and what was so very amazing was, that they perceived me! They were alive! There was communication. I was aware that they formed everything based on our thoughts... collective thought.

Needless to say, I was in awe. I remember thinking that everything is not real. I laughed to myself, it's real, but it isn't. It could be changed in a heartbeat if we only knew. I remember looking at the edges of the physical plane, it looked organic. The colors were so vivid and bold.

I was floating there focused on the scene when I felt a subtle presence. It began to grow, it grew so bright and so very strong, the energy was almost too much to bare. I began to feel very small in its presence, and with this tiny bit of hesitation, I let in fear. I started to feel fear of death. I wondered if I had killed myself. Suddenly, that tiny bit of fear in my soul grew out of control and consumed me like a cancer. I wanted desperately to live, to feel my body, my senses.

In this realm the rules were different. The five senses didn't apply. Communication was not with words, but with pictures and feelings. I felt like a fish out of water. I willed myself back, leaving this entity just as it had come upon me, in fear. I was finally able to anchor myself back into my body. As I sat up in bed I took note of the time. I had only been gone fifteen minutes, but it felt like days. On reflection, I knew who had come to see me. It was me! My higher self!

Behind The Veil   Art by Peter Nunnery   Gary Purviance

Big Foot - 1973 South Island, New Zealand. My friend Rick and I are asleep in a tent in the middle of a fern forest. The ferns are so dense that you can't see a white piece of paper in front of your face. I have a stomach ache. I get up to relieve myself. I'm wearing only underwear and it's cold. I'm leaning on a small sapling when suddenly my whole world is shattered by the sound of something huge smashing its way towards our camp.

There are no other people within miles of us. We have a 22-caliber gun, but it's in the tent. I can't believe Rick hasn't heard the sounds and come out blasting. I shudder in fear as I begin to hear its huge feet shake the forest floor as it enters our camp area.

I've never felt so helpless before. The fear is so all encompassing that I can barely think. I then hear the low loud snorts begin. It begins to mess around with our backpacks. I hear zippers being undone. It dawns on me that it has to be pretty intelligent to do that. I know that there are no large mammals on New Zealand, except sheep. After about two minutes it moves off into the woods. I hurry back into the tent and shake Rick over and over. He is dead to the world.

Eventually I find sleep. I awake to Rick shaking me and nearly screaming "Corey, wake up damn-it." It seems he couldn't wake me either.

"Bigfoot" we both scream at each other. His encounter was similar to mine, but there were two creatures and they crouched right next to the tent and snorted repeatedly right were our heads were. Rick was so scared he couldn't even move. The gun was on his mind, but useless as his body wouldn't respond. The creatures soon left and he drifted to sleep after trying to wake me.

The following morning we imagined ourselves the ones who would find and photograph them. I had a nice 35mm camera with me. We ate breakfast, loading up daypacks and took off in the direction they went. No footprints were to be seen as the forest floors are like sponges there.

About two hours later we stopped for lunch because we were both starving. It should have been around 11:00 am. I looked at my watch. It was 5:00 pm. I can't wait to find out what happened during those hours. We never saw a thing, at least consciously.

Big Foot

Corey Wolfe

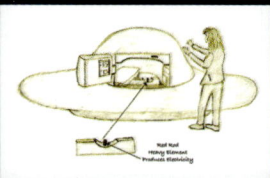

Helen - The year of the event is still uncertain, but I believe that sometime during the early 1970s is a likely date... I hope to have regression hypnosis to learn more.

The location was on the riverbanks of the river Ouse, a mile or so north of York City, Yorkshire, England. As I stood alone gazing at the mildly choppy waters of the river, I felt a relaxing, soothing effect from watching the flow of the water.

After a moment or two of standing in this relaxed state, I heard the voice of what I thought was a woman, say, "Hello," and indeed, as I turned to look to my right, there before me stood the image of a young female. She had dark hair, and was dressed in a navy blue, two-piece jacket and trousers, and her age seemed in her mid- twenties.

The woman approached and stood beside me and while starring me in the eye said, "Hello, I am a robot." Upon hearing this I turned my head to stare once again at the river and in a soft voice I said, "Oh, I see." The woman replied, "You don't believe

me?" in an angry tone. I answered, "Well to be honest, it is very difficult to believe." The woman replied, "I see. Yes I will accept that."

We got to talking and I learned many things from the robot, who I will nickname "Helen" because I cannot remember her real name. She showed me her strength by breaking pieces of metal, she told me about life in the galaxy and how the Earth was known about by most of the neighboring inhabitants.

Eventually, after walking some distance northward, Helen took me to a wooded area where she showed me her small black spaceship. Helen talked freely about how the spaceship worked and drew with her hands above the ship, the electrical diagram for a hyperspace generator.

Perhaps Helen talked freely about things because she was a lone robot, she had no masters or perhaps she escaped from one. I often wonder what happened to her. Unfortunately, she used some sort of mind control device to wipe away most of the memories of my encounter with her.

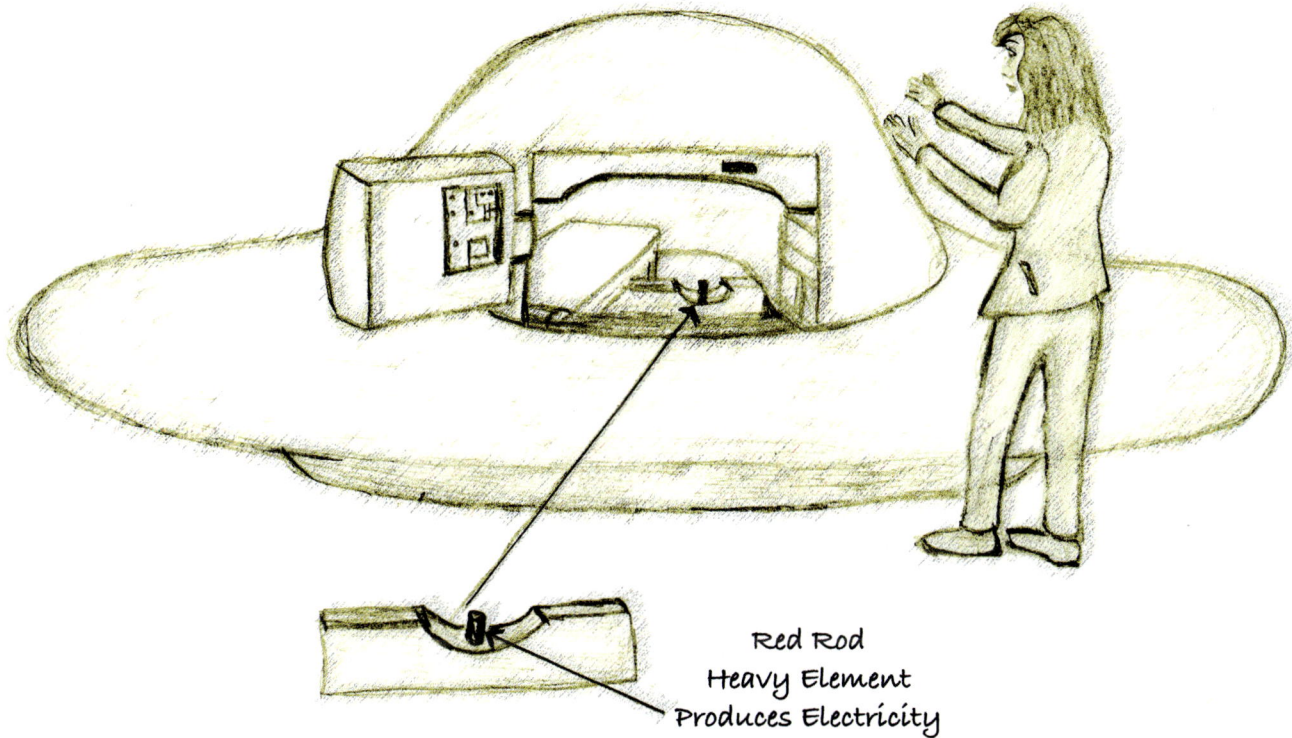

Red Rod
Heavy Element
Produces Electricity

Helen

Philip John Edwards

Future Home - I was wondering one day, where we would go if we destroyed the Earth. This is the image I received.

Portals are in the shape of pyramids or Ankhs. The Ankhs have the letters "I" and "S". I guess this is to indicate the IS. Perhaps their name for God. I doubt very much whether they would use English letters, maybe it's just my subconscious using a symbol.

I was told that imagining this Ankh in your mind's eye would pull you into another dimension. I tried it only once. It was so powerful a pull, it was amazing.

This new planet is much closer to the center of our galaxy as indicated by the densely packed stars on the left. Ships can be seen coming and going between dimensions.

Future Home   Corey Wolfe

Angel Whispers - I was a journalism major at KU, Lawrence, KS. back in the 60s, but I remember this life changing night as if it was yesterday. It was an incredibly clear, warm spring night. I had just won an award for sports broadcasting and I decided to make the short trip back to see the folks in Kansas City.

I was driving a blue 1966 Volkswagen, primarily to save money on gas. As I was heading toward the Kansas Turnpike I had a 'hunch' to fill up with gas. It wasn't one of those cloud parting, deep booming voices commanding me to fill up. It was just a gentle impression, like a key word: "gas." After I filled up, I was getting my toll ticket at the turnpike entrance, I had another 'hunch'. This time the message was 'seat belt', something I rarely did in those days. It's kind of like an invisible friend suggests, 'hey, buckle up,' yet your conscious interpretation says, 'I might as well fasten my seat belt'. I was obedient to the thought and buckled up as I headed eastward to KC.

Seven miles later, I came over the crest of a hill to see a dog lying in the middle of the turnpike. It startled me, because I couldn't tell if it was dead or alive. I moved to the right to avoid it and very quickly lost control of the car. The next thing I knew I was sliding sideways down the east bound lane. Quickly, I turned into the skid to make a correction. That maneuver put me into the grassy median dividing the east and west bound lanes. I jerked back to the right to get back onto the eastbound lanes. Bam!

Crash, glass breaking... the car rolled as it came out of the median. My head was literally inches above the pavement, waves of orange, red and yellow sparks surrounding me. Honestly I thought "this is it." But then an amazing calm gently surrounded me. I was at peace with what was to come.

I continued more than 165 feet across both lanes, finally coming to rest on the right hand service lane. The car had rolled back onto the driver's side when the car finally stopped. I was OK. Fortunately, I had neither hit anything, nor had anything hit me.

When I climbed out of the passenger door standing there alone, gathering my thoughts of what had just occurred and how lucky I really was I turned to my left, as if talking to someone standing next to me. I said out loud, 'Hey, Thanks. Thanks, a lot!' Now, physically there wasn't anyone standing there. Spiritually, I knew there was. At that moment, I knew I was not alone. If you've been helped in a challenging moment, you know what I mean.

When the highway patrolman was writing up the report, he said, "Did you have a full tank of gas?" "Yeah, I just filled up?" "And you were wearing your seat belt?" "Yeah. I don't normally, but I had a hunch to fasten-up tonight." "Well that hunch saved your life."

Eight years later during a personal consultation, I experienced a technique that gave me a direct, physical introduction to my team of angels. We all have them. You experience chills when you're relaxed and they're around. They are the source of peaceful, loving feelings, impressions and hunches that work in your best interest when you trust them. That night, those hunches indeed saved my life.

Angel Whispers   Art by Kim Carlsberg   William Hamilton

323

Dad's Birthday - Acrylic on canvas, 2009. This painting represents a major change in my life, and probably in the lives of many other people because we are at that kind of time on this planet now.

We've been talking about, and expecting major consciousness changes since the 80s, probably much earlier for some. With 2012 right around the corner, however, it seems like everyone I know, including myself, is in a state of flux that involves everything from "returning home" to relationship and career decisions. Mother Earth herself is at a dramatic turning point given her extremely fragile and volatile environment at this time as well.

While this artwork is not so much extraterrestrial in nature, it originates from the place where those other-worldly transmissions began, at home in Vancouver, British Columbia. It represents the culmination of twenty years of research, writing, lecturing, and painting about UFO contacts and the messages of the tall blue beings. It is, in a sense, where I have "landed."

This painting was begun on January 4th, 2009. It would have been my father's 77th birthday, had he not died in the 7th month of the year, 2008, 7 years after my mother passed away (also in July) from the same illness, lung cancer. Much of the teaching I received over the years from my guides was about the inevitable course our human lives take, and how to think, act, and develop oneself on the way to an enlightened mind, responding gracefully when our transition time arrives.

Many people say it sounds like I am in contact with angels, but in true form, they are clusters of energy, propelled by streams of vrili, and contain all of the intelligence of the Universe. Their language is sound, color, and when in the presence of a human, can take the form of scents like roses or jasmine.

In such a state, they may appear as balls of light that are "ships" in the sky. An early group of paintings I did was called "The Spectrum Light Vehicles", and depicted the ships as they manifested in my visions, frequently in groups of three, forming a triangle.

In the moments after my dad passed away, quietly and peacefully at home on his favorite couch, I heard a bell ring loud and clear, in the same room. It rang a second time, and my sister-in-law heard it... then a third and my brother, who had been on the phone, heard it. We searched the room and found a small brass bell that had the same tone as the one the three of us had just witnessed.

It had a profound effect on us in the moments following the death of a loved one. My brother stayed up most of the night researching and finally found that we had literally, and without knowledge, had participated in an ancient Japanese Buddhist, "Three Teacup Ceremony" for the newly deceased that allows the person smooth passage to Nirvana.

This painting, depicts a "re-birth", the surrounding energy, and the portals of time and dimension through which we travel... extraterrestrial, angel, and human.

Dad's Birthday

Susan Gordon

Gotcha - 6 -1972. I was seven, at my Oma's, playing soccer with the locals till the kid with the soccer ball had to go home. The rest of us still wanted to play, so we made our shirts into a ball. When it was my turn to kick, the ball unraveled and my shirt blew onto my Oma's roof. I ran up the stairs and onto the gravel roof where the shirt sat on the very edge. I then slowly, baby-stepped over to it, where twenty five feet below, my friends cheered me on. I took a step back and kicked so hard it threw my body forward and I plummeted face first, two stories, towards solid concrete.

Everything around me immediately slowed to a near standstill, but my mind was clear. Millions of thoughts raced through me but I had time for every last detail. I saw the ground and the neighbors ugly orange car parked one space over from where I would impact. I lingered in the air for many minutes watching my thoughts flash faster and faster, as the world moved slower and slower.

Suddenly I was aware of small, warm hands gently wrapping around my ankles. An overwhelming feeling of calm and trust came over me for whomever was holding my feet. I caught a glimpse of a small being the size of a 5 year old boy, dressed in white, but I particularly noticed his huge, Elfish, green eyes. I thought how strong he was to hold me up like that. I had no sensation of falling, it was as if I was being lowered to the ground. Within four feet of the ground, I got very tired. I said "thank you" and felt a warm rush of goose bumps as he replied "you are welcome." I closed my eyes and drifted

I was back in my mother's womb, experiencing my own creation where any thoughts other than love, did not exist. I was floating in a warm space like a swimming pool, but I was able to breath. My awareness shifted and I was in the center of the cosmos. My mother's womb and the center of the universe were the same place. I had no concepts apart from unconditional love, peace and contentment, and I wanted to stay in that place forever.

After what seemed like eternity, I heard distance bells from a church tower. I could see nothing, I could only feel and hear. Though I was only seven, I realized I was experiencing my birth into this world once again. I felt absolute joy and connection to the life process.

A muffled human scream jarred me back into self awareness. The rush of sensory information slammed into my quiet mind, like a crashing fright train. I did not open my eyes, for at that moment I realized I had been separated from my body and was merging back with it. It wasn't until I heard my name that my full identity returned and the inner peace began to fade. I didn't want to wake up: like rousing from a fantastic dream, I wanted to go back to sleep and re-enter the rhapsody. I became aware of my body again from the pressure of the cold concrete beneath me and the afternoon breeze upon my face. I suddenly realized that I felt no pain, in fact, I felt absolutely great.

A throng of people had been towering over me crying with wracking sobs. When I finally opened my eyes and took a deep breath, a unified gasp ensued and with shocked looks, the entire crowd retracted. A moment later, they lunged towards me smothered me with hugs, crying even harder. I remember thinking, "What's the big deal?" The parents and neighbors later revealed the big deal... I had been "gone" for at least seven minutes!

Gotcha

Roy Young

Howard - I heard this story from both my grandmother and my grandfather, first when I was about eleven or twelve. My grandmother cried when she told it to me the first time, apparently soon after the event, which would have been around 1964 or 1965.

My grandparents lived on Connecticut Avenue, here in Knoxville, TN, a small neighborhood across from the recess field of Lonsdale Elementary School and two doors down from the Methodist Church and just a block from the bus stop to downtown. It was an area on the way down economically, but struggling to maintain itself above the poverty level. The small, red-sided house in which they lived was surrounded by a low, chain-linked fence, placed there by my grandfather mostly to keep rougher neighborhood kids from wandering through.

On this particular day, my grandfather was in the back of the house and my grandmother was sitting in the living room watching one of "her stories", as she called daytime dramas. She heard the metallic chink of the fence open and close on the front walkway to the porch, footsteps, then a quiet knock on the door. She opened the door and there stood a boy she guessed to be about four years old, disheveled dark brown hair, worn "Levi's" and a flannel shirt - a bit too large for him. He just stared at her, sadly, she thought, until she spoke and said, "Are you lost?" "No, Ma'am," said the little boy. At that moment, my grandmother's eyes began to be teary as she told me the story. She said, "I felt cold. And I'm ashamed to say I felt scared and didn't know why. Where do you belong?" my grandmother asked.

"I belong here." The little boy replied. My grandmother asked the child to wait a moment while she went and got her husband, my grandfather. When they returned to the door, the boy was gone. That was on a Saturday. My grandmother waited with my grandfather the next day at the time the boy had come, but he did not come back, not then or the next day, or the next. My grandfather was a bit irritated at my grandmother for becoming so upset at what was obviously one of the neighborhood kids coming to the door for a glass of water or something to eat, but my grandmother insisted that it was something else. She demanded that my grandfather stay home the following Saturday.

So it was that afternoon when the little front gate clinked, they were both sitting in the living room quietly, not really talking, just sitting in that quiet way that dumbfounds us as children. When the steps were heard on the front porch, my grandfather motioned, "I'll get it," and rose to answer the door. My grandmother was peering from right behind him at the first knock, her small frame almost completely hidden by his tall stout one. Tall and stout he might be, but his was a voice of kindness to many children in the neighborhood. Again it was the same solemn little boy, dressed the same way. Again, he said nothing. "Do you live around here?" my grandfather asked gently after a moment. "Yes, sir," said the boy. "Where do you live?" asked my grandfather. "With you." And with that the boy turned and started down the porch steps. "Wait!" cried my grandfather, ignoring my grandmother leaning against his back. "Who are you? What's your name?" The boy clinked the gate open and looked back. "Howard." He turned onto the sidewalk and, as they both watched, disappeared between one step and the next.

You see, Howard was the name of their first son, born with spina bifida in the early 1910s. He lived only a few weeks.

Howard   Art by Kim Carlsberg   Chuck Hagaman

7th Deja Vu - This story took place in 1984 during the summer Olympics. My girlfriend Sheri and I sat on my bed discussing metaphysics.

I had a Deja Vu. This feeling flowed into a second one, and then a third Deja Vu. At this point 'reality' was beginning to shift. Enough of me was still here to grab Sheri's hand and say to her, "Something weird is happening, please don't say anything for a while, and see if you can go with me." The adventure had begun.

The Deja Vu feelings continued to come. They were different than most I had felt in the past. They had nothing to do with a past event that I was now remembering. They were happening as a means to elevate my consciousness.

As the seventh one hit me I began to feel that always- welcome-sensation of being in touch with the All. My consciousness was above my head, and in a very clear state. Then to my surprise, I found myself standing on stone stairs. This wasn't my imagination, this felt real, yet different. I gazed down, and seemed to be high on a mountain range. There were clouds below my position.

I looked up, and about six steps above I saw a cave entrance. A man dressed in white robes stepped out of the cave and said to me, "Congratulations, today you made your quota." At this I was thoroughly amazed. I seemed to know what he was talking about. Earlier in the day I had turned another person on to meta-physics. I had spent the last sixteen years doing this at every opportunity. If something works for you, you try to share it. I replied to him, "But I'm so young." To this he had no reply. I then asked if Sheri could feel all that was going on and he said that she was not yet ready.

This teacher told me that I had to conquer all fears in order to hurry my evolution along, "Knowledge lies where fear is not." We examined my greatest fear at the time, one of letting go of control. I was afraid to surrender to **the All** completely for fear of losing my individuality.

He said that I would soon have the opportunity to face this challenge. He gave me a moment to collect my thoughts. 'Time' is different 'there', and much was exchanged in a short while.

What happened next became a very big lesson in my life, one I use to this day. His next words were, "You may now come home, or stay." My thoughts came fast and furious. Did he really mean what I thought he meant? Go home? Die?

I asked him what Sheri would do if my body actually died. I had barely got that thought out, when in a flash I found myself back on my bed. The 'vision' was over. I had been dismissed, I had hesitated... doubted.

7th Deja Vu

Corey Wolfe

God Bubbles - I will never hesitate again! This is the lesson I have learned. If ever you get an opportunity to gain wisdom, go for it! If you hesitate long enough to allow the left-brain to get into the picture and start analyzing the situation, the odds are you'll chicken-out.

Higher reality has nothing to do with the analytical brain. It has everything to do with intuition.

As he said within days I had the opportunity to test my newfound lessons and myself. I had lain down to take a nap and had just closed my eyes, when I saw 'The Door of Fear' in my mind's eye.

There is a place deep within all of us that we are afraid to enter because all the demons and fears we hold are within. I had often seen this portent of doom while meditating, but always ran from it.

Now I had to change all that. Using the lessons just learned I didn't hesitate; I just mentally closed my eyes and ran as fast as I could straight toward the door, and jumped!

What I found behind the door of fear was the last thing I had expected. I found God! The place where there is no difference between a question and its

answer. I learned that it is the door itself that we fear. The door is a manifestation and representation of our own imagination. I lovingly invite you to join me, in doing our best, to imagine only positive things and perhaps such doors will dissipate. It may save you a jump.

I later had an opportunity to learn about surrender. While meditating one day an idea came to me. Why not try to mentally erase everything thing that had to do with myself. I began by ignoring anything that came to mind that had anything to do with my personality or my wishes.

I found as the meditation went deeper that there was a new sense of individuality arising, something I had never experienced before. It proceeded quite naturally as if I had done this many times previously. I found 'myself' racing over what looked like rolling mounds of deep velvety green at the altitude of one foot. A voice came to my ears and said, "You loose the self, but gain the Other."

Again, I was in the place where any question I could muster was simultaneously matched with its answer. I questioned the Other for hours and heard many wonderful and unexpected things, but that's an ***Other*** story.

God Bubbles   Corey Wolfe

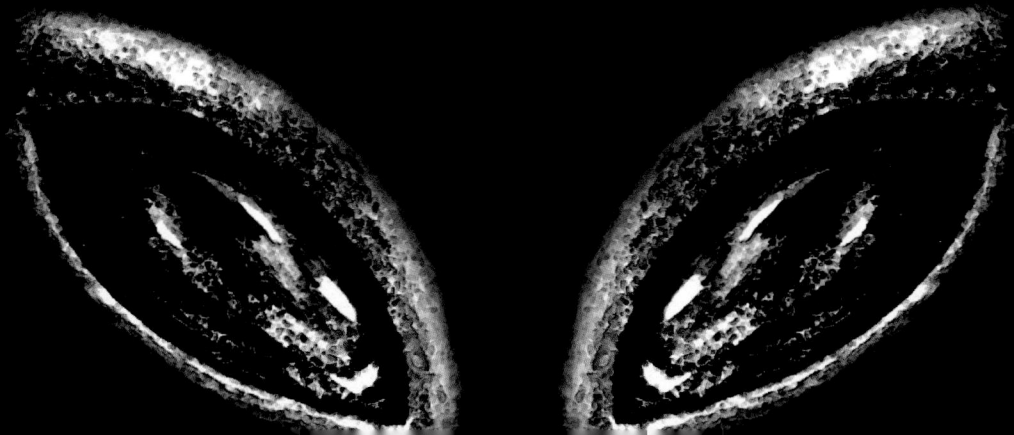

The Art Of Close Encounters

Encounter Art Conclusion...
the message is clear...
we are not alone...
our little blue planet is precious...

and it is time we live respectful
of these truths.

Blessings, Kim

About The Contributors

Usko Ahonen
Satamakatu 4 A 15 74100 Iisalmi Finland
 Story:
 -The Space Aliens

Margarita (Rita) Andreeva
randreeva.imagekind.com
scribd.com/ritaandreeva
myspace.com/ritaandreeva
youtube.com/ritaandreeva
facebook.com/randreeva
 Story and Art:
 -Alien and Human
 -Alien

Darryl Anka
www.bashar.org
 Art:
 -Beyond My Wildest Dreams
 -In The Blink Of An Eye
 Story:
 -Bashar's Craft

Laura Barbosa
Fine Artist (NJ Sole Proprietorship)
Location: Online Art Studio. Websites: barbosaart.
etsy.com, artprints.imagekind.com. Other: The Heart
Of Art Blog: laurabarbosa.wordpress.com. Laura's
Public Collection at the Community Medical Center in
Toms River, NJ Raised $7500.00 for a local therapy
dog organization (3 Original Paintings) Laura's
artwork is currently licensed with Icon Shoes.
www.iconshoes.com
 Story and Art:
 -Shape Shifter

Dalton Bagby
 Story and Art:
 - Paradigm Shift

Tim Beeken
blogger.com -__beeks-purity.blogspot.com
 Story and Art:
 -Through The Woods
 -A Saturday Afternoon

Brent Berry
Brent Berry Arts
http://brentberryarts.com
 -Story and Art:
 -Saucer Surfing
 -Unsuspecting
 -The Gathering
 -Beaming Horseplay
 -Rush Hour
 -No Where To Hide

Mike Bird
 Story and Art:
 -The Little Monks

Kim Carlsberg
www.kimcarlsberg.com
http://kimcarlsberg.imagekind.com
www.outtherezone.com
www.theartofcloseencounters.com
www.closeencounterspublishing.com
www.beyondmywildestdreamsbook.com
www.facebook.com/kim.carlsberg

Art:
-Seeing The Light
-While You Were Sleeping
-Through The Eyes Of A Child
-Bronze Beauty
-Vasquez Visitor
-Shining Brightly
-Silver Sighting
-Burning
-Overpass
-Bashar's Craft
-The Guiding Light
-The Alien Hunter
-The Space Aliens
-Strange Visitors
-Alien Jewel
-Children Of The Phoenix Lights
-Fairy Presence
-Crop Circles
-Angel Whispers
-Howard
Story:
-Beyond My Wildest Dreams
-In The Blink Of An Eye

Sanni Ceto
845palmerstreetdeltaco-81416
        Story and Art:
        -Mind Scan

David W. Chace
www.conceptart.org/forums
/showthread.php?t=136530
        Story and Art:
        -Galactic Family

-Reptilian Types
-Renjeck
-Reptilian 1
-Reptilian Body
-Reptilian Gray
-Communion
-Old Gray
-Military Hybrid
-Yurani
-Estartleah
-Blue
-Mantis
-Glowing Goblin
-Alien Hands

Chuck Chroma
Chuck Chroma LLC
4820 W. 36th Ave.
Denver, Colorado 80212
www.chuckchroma.com
        Story and Art:
        - Making A Point
        - The Great Room
        - Odd Aunt Shirley

Mike Clelland
http://hiddenexperience.blogspot.com/
        Story and Art:
        -Bedroom Window

Jolibeth Cope:
        Story:
        -Bronze Beauty
        -Silver Sighting

Ryan Cope
  Story:
  -Vasquez Visitor

Bryan DeFlores
LightQuest
PO BOX 93658 Las Vegas, NV 89193
www.bryandeflores.com
  Story and Art:
  -Through The Looking Glass
  -Blue Starfire
  -Money Tree

Raven De La Croix <aka: De Lumiere>
"Ravens Cosmic Portal" - P.O. Box 4792 -
Sedona, AZ 86340
www.RavensCosmicPortal.com
www.RantingsOfaMadWoman.com
www.RavenDeLaCroix.net
  Story and Art:
  -Abduction Seduction

Miriam Delicado
#325 1027 Davie Street
Vancouver BC Canada V6E-4L2
www.bluestarprophecy.com, www.alienbluestar.com,
miriam@bluestarprophecy.com. Author: Blue Star
Fulfilling Prophecy. I have a blog and forum on my
website dealing with UFO's and positive ET contacts.
The focus is on the positive contacts and messages
from the "Tall Blonds." The 'Great Gathering' is what
the Blond's goal is: to unite humanity in order to live
a more balanced life with ourselves and the earth.
  Story:
  -Nordic and Friends

Christine "Kesara" Dennett
http://kesara.weebly.com/extreterrestrials.html
kesaraart@gmail.com
  Story and Art:
  -Eliad
  -Zetas
  -Hybrid and Family
  -Mantis
  -Praying Mantis w Triangle Pin
  -Oldest
  -Green Reptilian
  -Nordic Female
  -Nordic Male

Eric Rojo Dotel
(son of Albert Rosales)
  Art:
  - Blackout Days
  - Winged Creature

Philip John Edwards
www.aliendiscoveries.blogspot.com
  Story and Art:
  -Helen

Christian Fedor Flores Cordova
www.facebook.com/#!
/profile.php?id=598023576
  Story and Art:
  - Tall Whites

DW Freeman
RNABrand
www.rnabrand.com
  Art:
  - Probes Happen

Charla Gene
5342 N. Paseo de la Terraza, Tucson, Az 85750
http://enchantresse.imagekind.com/
        Story and Art:
        - Celestrial Celebration

Monica Geyer
http://lyra77.blogspot.com/
www.facebook.com/mstarship
        Story and Art:
        - The Innocents

Susan Gordon
Green Pony Productions, Vancouver, B.C.
susan.greenpony@gmail.com
        Story and Art:
        - Blue Beings
        - Dad's Birthday

Mike James Gorman
6 Manchester Road, Leigh, Lancashire, UK
pilsky@hotmail.co.uk
        Story:
        -Burning

Susaye Greene - Ambassador to California and
the World for the Goodwill Treaty for World Peace
(www.goodwilltreaty.org) Upcoming proiects: New
cd collaboration with UK artist Matti Roots, James
Bartlett's indie Film "Nostalgia"... I did the poster
and sang on the soundtrack.websites:
www.supremextreme.com
www.supremextreme.deviantart.com
www.supremextreme.imagekind.com
www.myspace.com/susaye
        Story and Art:

Rosemary Ellen Guiley
137 Danbury Rd. PMB
336 New Milford CT 06776
www.visionaryliving.com
www.djinnuniverse.com
        Story:
        -Fairy Presence

Barbara Gunderson
Flwrchld90@aol.com
        Story and Art:
        -Touched

Chuck Hagaman
        Story:
        -Howard

William Hamilton
willhamilton07@gmail.com
        Story:
        -Angel Whispers

Cliff Hare
Infohazard Gallery
www.infohazardgallery.com
infohazardgallery@yahoo.com
        Story and Art:
        -Alien Tourists In Hell

Angela Vaughn Hausinger_R.M.
Universal Energies Therapies, Healing
Hands, Spiritual Counceling. Huston Texas.
www.myspace.com/avh-caapi
LookingToTheStars@hotmail.com
        Story:
        - Taken For A Ride

Shawn Kevin Jason
www.zetascafe.com
    Story and Art:
    - Area 51 Triangle

Akira Kawatech
    Story and Art:
    -Traveling

Elle Keith
www.childrenofthephoenixlights.com
www.facebook.com/ElleKeith
PERSONAL CHANNELED LETTERS FROM YOUR
HYBRID CHILDREN: Join us on our
Facebook fan page where these beautiful
letters are posted to their mothers and fathers.
If you feel or know you have hybrid children,
I will telempathically channel a letter for you
from your child... using automatic writing. I
am compiling letters into my book entitled,
LETTERS FROM CHILDREN OF THE PHOENIX
LIGHTS:BRIDGING GALACTIC FAMILIES. The
book also includes questions that will be asked
of the children about their lives. If you would
like to submit questions to ask the children,
they would love your participation. Contact
Elle at ellekeith@msn.com.
    Story:
    Children Of The Pheonix Lights

Danion Kelly
    Story and Art:
    - Dark ET
    - Inside The Ship

Kristi Kennington
Studio 88 Fine Art, Fort Worth Tx
www.studio88.org
    Story and Art:
    -Who Are The Aliens
    -El Receptor de Huevos
    -Thes Grises

Jacky Kozan: Contributor for Daniel C. and
Investigators: Georges Metz & Jean-Claude
Venturini. Published in the magazine: "Lumières
Dans La Nuit" Nr 377 (Feb. 2005).
    Story and Art:
    -Emerald Trio
Extracts from the report of Gerard Deforge published
in the Magazine "Lumières Dans La Nuit" Nr 352
    Story and Art:
    - Light Beams

Melissa Kriger
    Story and Art:
    -The Orange Orb

David Patrick Kuhlman
UFO Information for Human Rights,
113 Tom Bing Road Silver Creek, Georgia 30173
www.ufoforhumanrights.com
    Story:
    -Seeing The Light

Melinda Leslie
www.alienexperiencers.com/MelindaLeslie01
Public experiencer 20 yrs/UFO researcher 23 yrs.
Work featured in book Camouflage Through Limited
Disclosure: Deconstructing a Cover-up of the Extra-
terrestrial Presence by Randy Koppang. Lecturer:

MUFON, the X-Conference, Bay Area UFO Expo,
UFO Expo West, International UFO Congress, Whole
Life Expo, etc., TV & radio guest, Director UFO
lecture series 9 yrs hosting most prominent names
in UFOlogy.
Story:
- Detour

Lee Louden
Story:
-The Orange Sphere

Justin Lowe
ImageKind.com screen name:
alien_elf_art
Story and Art:
- Making Things Clear

Larry D. Lowe
Phoenix UFO Examiner
ufo.larrylowe.com
Story and Art:
- The Phoenix Lights

Karen Lyster
www.karenlyster.com
Story and Art:
-The Elder
-Little Miss Bright Eyes
-The Lightworker
-Aqua Eyes The Mystic

Tonia Madenford
www.screenaddiction.com
www.myspace.com/screen_addiction
www.imdb.com/name/nm1375697

www.facebook.com/screen.addiction
www.youtube.com/screenaddiction
Story:
- Reaching
- Rising

Stephen Martin
Martindoolittles Alien Doodles
www.aliendoodles.com
Story and Art:
- The Fear Egg
- Home World
- Wierdest Year
- Lizard Bloke

Elliott Maynard, Ph.D
Arcos Cielos Research Center,
P.O. Box 20069, Sedona, AZ 86341, USA
http://www.arcoscielos.com
Story and Art:
-Arcturan Matter Transmitter
-Lyran Dream Generator

Michael Austin Melton
425 Featherbed Lane, Glen Mills, PA 19342
paradigmshiftforward.blogspot.com
Story:
-Through The Eyes Of A Child

Robert D. Miles
8699 Silver Creek Drive
Show Low, AZ. 85901
www.safespaceproject.com
www.fastwalkers.com
rdm@safespaceproject.com
Story and Art:
-Shimmering Lights

Wayne Miller
10662 Rhodesia Ave. Sunland, Ca. 91040
WaynePBA@GMail.com
Story:
- Towering
Molly
Story and Art:
-The Puzzle

Jim Nichols
2801 W. Fresno St. #2
Tuscon, Az 85745-1745
www.jimnicholsufoart.com
Winner - 2 EBE Awards at 2010 IUFOC conference
- video doc; Nazi Saucers, "The Aldebaran Mystery.
"Noted artist, investigator, lecturer... UFO illustrations
published internationally in all print medias & routinely
featured on "The History Channel's "UFO Hunters." A
prolific writer... 21 BLOG essays on EXOPOLITICS - the
political and social impact of a possible extraterrestrial
reality and art book, "Case Files Revealed"... 40 untold
stories behind most dramatic UFO illustrations based
on actual accounts all available on website.
Stories and Art:
- Puerto Rico Encounter
- Cylinder Ship
- Semjase
- Country Road

Peter Nunnery
pnunnery@klm-creative.com
http://www.facebook.com/pete.nunnery
Art:
-Beyond The Veil
-Behind The Veil

Oilboy
www.flickr.com/photos
/33896608@N00/sets/
Story and Art:
-Monet Grays At Roswell
-Alien Eye At Roswell
-Clown Grays At Roswell

John Pagan
johnpagan@furyinthegarden.com
Story and Art:
-UFO Abducted?
-Dream Within A Dream

Andrew Pearce
DNA Designs
_dna-designs.co.uk
Story and Art:
- Around The Bend
Art:
- Towering

David Penfound
19 Beaumaris Close.Andover.
Hampshire.England.UK.SP10 2UB
david@davidpenfound.com
Story and Art:
-The Surf Is Out There

Garth Perfidian
G. Arthur (GARTH) Edwards - Artist of Paradox.
Lucky 7 Gallery & Gifts - 140 Rainier Ave S. - #7
Renton, WA. 98057
www.freewebs.com/garthfromseattle
www.freewebs.com/garthsdream

Story and Art:
-The Man Who Volunteered
-Whistler
-The Night Wolves
-Female Alien
-Saucer Down
Art:
-The Orange Sphere
-Taken For A Ride

Jeffery Pritchett
The Church Of Mabus Radio
www.mabusincarnate.com
    Story:
    - Sign Of The Times

Randy Pull
    Story:
    -The Guiding Light

Kevin Reid
    Story:
    -Overpass

Michael H. Rogers
    Art:
    -Fire In The Sky

Bruce Rolff
www.facebook.com/brucerolff
    Story and Art:
    - Off Shore Encounter

Albert Rosales
UFOINFO Humanoid Encounters
    Story:
    - Blackout Days
    - Winged Creature

Andy Russell
http://thetruthhides.wordpress.com
    Story:
    While You Were Sleeping
    Strange Visitors
    Crop Circles

Helen Sanderson
    Story and Art:
    Drakonian

Derrell Sims
Derrel W. Sims, R.H.A, CM.Ht.
P. O . Box 60944 Houston, Texas, 77205
www.AlienHunter.org
FATE MAGAZINE, BLOGGER,
Derrel W. Sims, aka, TheAlienHunter
    Story:
    Alien Hunter

Rick Smith
www.ufoteacher.com
www.alienalley.com
    Story and Art:
    - Laboratory

Yuichi Tanabe
Mistic Media
www.misticmedia.com
  Story and Art:
  -Exploring The Coastline
  -Moai Small Planet

Paul Van Scott aka Dr. Paxton
www.FinalScoreProducts.com
  Story and Art:
  -Alien Face
  -Area 51

Travis Walton
www.travis-walton.com.
Experiencencer/author of Fire in the Sky; true story
of alien abduction in nothern Arizona in 1975. The
movie of the same name was released by Paramount
Pictures in 1993. Updated version of new book due in
2011.
  Story:
  -Fire In The Sky

Albert Lloyd Williams
www.albertlloydwilliams.com
oldrecluse.blogspot.com
  Story:
  -Shining Brightly

Corey Wolfe
12014 N.E. 192nd Ave.
Brush Prairie, Wa 98606
www.coreywolfe.com
  Story and Art:
  -Pretty Feet
  Gray Salute
  -The Is

-Inside A Very Big Ship
-Future Home
-Seven
-Big Foot
-7th DejaVu
-God Bubbles
Art:
-Nordic and Friends
-Slanted Eyes

Dean Wolfe
2120 W. 240th St., Lomita, CA 90717
  Art:
  -Detour

Emma Woods
www.ufoalienabductee.com
emma@ufoalienabductee.com
  Story and Art:
  -Torn

Roy Young
Young Productions
6606 W. Prickly Pear T. Phoenix, Az 85083
  Art:
  -Reaching
  -Rising
  Story and Art:
  -Gotcha

Dale Ziemianski,
9471 Buckeye Troxel Rd,
Sugar Grove, OH 43155
  Story and Art:
  -Adam (Back Cover)
  -The Prank

PAGE INDEX -